SMP interact 7T

Teacher's guide to Book 7T

CAMBRIDGE
UNIVERSITY PRESS

The following people contributed to the writing of the SMP Interact key stage 3 materials.

Ben Alldred	Ian Edney	John Ling	Susan Shilton
Juliette Baldwin	Steve Feller	Carole Martin	Caroline Starkey
Simon Baxter	Rose Flower	Peter Moody	Liz Stewart
Gill Beeney	John Gardiner	Lorna Mulhern	Pam Turner
Roger Beeney	Bob Hartman	Mary Pardoe	Biff Vernon
Roger Bentote	Spencer Instone	Peter Ransom	Jo Waddingham
Sue Briggs	Liz Jackson	Paul Scruton	Nigel Webb
David Cassell	Pamela Leon	Richard Sharpe	Heather West

Others, too numerous to mention individually, gave valuable advice, particularly by commenting on and trialling draft materials.

Editorial team	Project administrator	Design	Project support
David Cassell	Ann White	Pamela Alford	Carol Cole
Spencer Instone		Melanie Bull	Pam Keetch
John Ling		Nicky Lake	Jane Seaton
Paul Scruton		Tiffany Passmore	Cathy Syred
Susan Shilton		Martin Smith	
Caroline Starkey			
Heather West			

Special thanks go to Colin Goldsmith.

CAMBRIDGE UNIVERSITY PRESS
Cambridge, New York, Melbourne, Madrid, Cape Town, Singapore, São Paulo

Cambridge University Press
The Edinburgh Building, Cambridge CB2 8RU, UK

www.cambridge.org
Information on this title: www.cambridge.org/9780521537988

© The School Mathematics Project 2003

First published 2003
5th printing 2007

Printed in the United Kingdom at the University Press, Cambridge

A catalogue record for this publication is available from the British Library

ISBN 978-0-521-53798-8 paperback

Typesetting and technical illustrations by The School Mathematics Project
Illustrations on pages 39, 40 and 162 by Chris Evans
Photographs by Graham Portlock
Cover image Getty Images/Randy Allbritton
Cover design by Angela Ashton

Contents

Introduction

Teaching approaches

SMP Interact sets out to help teachers use a variety of teaching approaches in order to stimulate pupils and foster their understanding and enjoyment of mathematics.

A central place is given to discussion and other interactive work. In this respect and others the material supports the methodology of the *Framework for teaching mathematics*. Questions that promote effective discussion and activities well suited to group work occur throughout the material.

Some activities, mostly where a new idea or technique is introduced, are described only in the teacher's guide. (These are indicated in the pupils' book by a solid marginal strip – see below.)

Materials

There are three series in key stage 3: *Books 7T–9T* cover up to national curriculum level 5; *7S–9S* go up to level 6; *7C–9C* go up to level 7, though schools have successfully prepared pupils for level 8 with them, drawing lightly on extra topics from early in the *SMP Interact* GCSE course. Integrated carefully into *Book 7T* is material covering the government's 'essential teaching objectives in a year 7 catch-up programme'.

The year 7 books share much common material – a benefit where mixed attainment groups or broad setting are used for an initial settling-in period, or where the school covers topics in parallel to ease transfer between sets. To help you with your planning, links to common and related material – between *Book 7T* and *7S*, and between *7S* and *7C* – are shown to the right of unit headings (in both the pupils' book and the teacher's guide); for example, unit 2 of *Book 7S* has the links '7T/5, 7C/2', meaning there is common or related material in unit 5 of the less demanding *Book 7T* and in unit 2 of the more demanding *Book 7C*.

All three year 7 books start with a collection of activities called 'First bites', designed to help you get to know pupils and to give them an enjoyable and confident start in mathematics at secondary school.

Each of the three *T* books contains a collection of 'Number bites' – short activities for regular practice of number skills.

Pupils' books

Each unit of work begins with a statement of learning objectives and most units end with questions for self-assessment.

Teacher-led activities that are described in the teacher's guide are denoted by a solid marginal strip in both the pupils' book and the teacher's guide. Some other activities that are expected to need teacher support are marked by a broken strip.

Where the writers have particular classroom organisation in mind (for example working in pairs or groups), this is stated in the pupils' book.

Resource sheets

Resource sheets, some essential and some optional, are linked to some activities in the books.

Practice booklets

For each book there is a practice booklet containing further questions unit by unit. These booklets are particularly suitable for homework.

Teacher's guides

For each unit, there is usually an overview, details of any essential or optional equipment, including resource sheets, and the practice booklet page references, followed by guidance that includes detailed descriptions of teacher-led activities, advice on difficult ideas and comments from teachers who trialled the material.

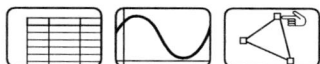

There is scope to use computers and graphic calculators throughout the material. These symbols mark specific opportunities to use a spreadsheet, graph plotter and dynamic geometry software respectively.

Answers to questions in the pupils' book and the practice booklet follow the guidance. For reasons of economy answers to resource sheets that pupils write on are not always given in the teacher's guide; they can of course be written on a spare copy of the sheet.

Assessment

Two short general assessments ('Can you …? A' and 'Can you …? B') are included as sheets 3–6 in the resource sheet master pack to help you identify, early in year 7, pupils with difficulties in some areas of number. See page 19 for further details.

Unit by unit assessment tests are available both as hard copy and as editable files on CD (details are at www.smpmaths.org.uk). The practice booklets are also suitable as an assessment resource.

Mental methods and starters

Mental methods

This section sets out a teaching programme covering the following mental skills:

- addition and subtraction of up to two-digit numbers
- subtracting a two-digit number from 100 (and working out change from £1, £5, etc.)
- doubling and halving
- multiplying a two-digit number by a single digit

Work on developing mental calculations methods needs to be spread out on a 'little and often' basis – and the lesson starter is ideal for this – but with perhaps an occasional longer session focusing on specific skills.

Pick up on pupils' own ideas and methods and discuss and compare them. Pupils will be more forthcoming in describing their methods if they do so in pairs or small groups.

Often the method will vary with the numbers involved. For example, $25 + 19$ might be done as $25 + 20 - 1$, whereas $25 + 26$ might be done as $25 + 20 + 6$, or even $2 \times 25 + 1$.

Pupils who have their own preferred methods will still benefit from seeing and using others. But in the end it is accuracy and efficiency which count, so pupils should not be forced to learn a particular method when they have others which work for them.

Each method is described very briefly below. However, be prepared to spend a fair amount of time working together as a class on examples of each method. (A few examples of suitable calculations are given to save you having to think them up on the spot!)

Vary the language you use. For example, in the case of addition you can use 'add', 'total', 'add to', 'add together', 'increase by', '… more than', etc. Ask some questions in context.

Adding

An **unmarked number line** is very useful model, though to start with you could mark just the tens graduations. Draw the diagrams on the board.

Preliminary stage

Make sure at the outset that pupils can easily add a single-digit number or 10. A good way to do this is to practise counting on in steps.

Start by drawing an unmarked number line and put a number on it, for example 45. Get the class to count on together in 2s, 3s, 4s etc. and 10s. Do the same with other starting numbers.

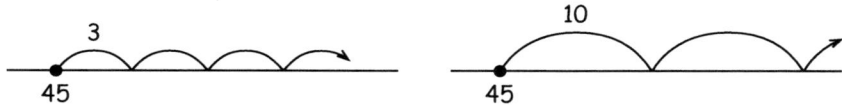

Practise adding a multiple of 10 (for example 47 + 30) in one jump.

Adding on tens then units, with no 'carry'

Example 45 + 13   Where are you after you have moved forward 10?

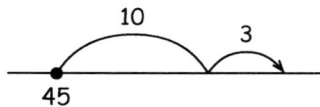

Example 26 + 23   Where are you after you have moved forward 20?

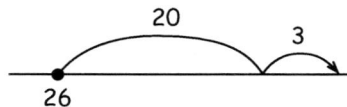

Other examples   83 + 12   46 + 42   61 + 38   74 + 23   53 + 46   41 + 55

Adding 9s by adding 10s then adjusting

Examples 45 + 19   36 + 29

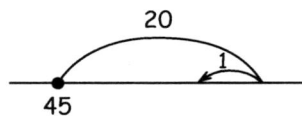

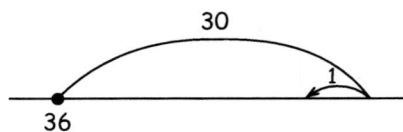

Where does the first jump take you?

Other examples   52 + 19   23 + 39   29 + 44   59 + 28   64 + 29   39 + 47

Adding the nearest multiple of 10 then adjusting

Examples 44 + 38   47 + 56

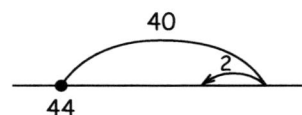

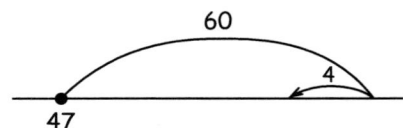

Other examples   23 + 18   37 + 28   35 + 37   47 + 26   38 + 43   25 + 47

Adding on tens then units, with 'carry'

Examples 45 + 28   76 + 65

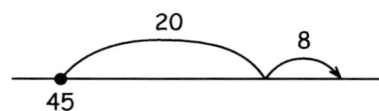

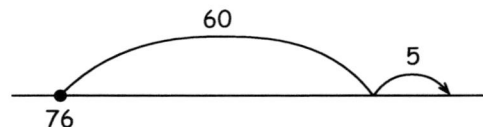

Other examples   39 + 12   47 + 24   36 + 56   37 + 25   55 + 17   38 + 32

Subtracting

As with addition, an unmarked number line is very useful.

Preliminary stage

Make sure pupils can easily subtract a single-digit number or 10.

Start by drawing an unmarked number line and put a number on it, for example 63. Get the class to count back together in 2s, 3s, 4s etc. and 10s. Do the same with different starting numbers and practise counting back.

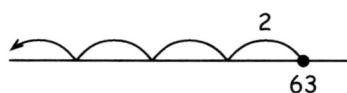

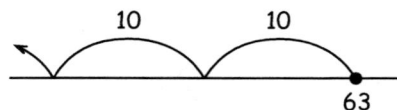

Practise subtracting a multiple of 10 (for example, 63 – 40) in one jump.

Counting on when the numbers are close

Examples 82 – 79   104 – 98

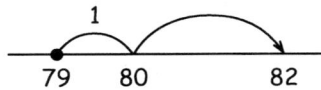 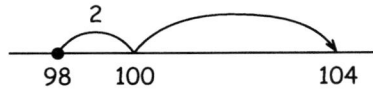

Other examples 35 – 32   41 – 39   72 – 68   90 – 87   53 – 49   71 – 67

Subtracting to a multiple of 10 then adjusting

Example 22 – 7

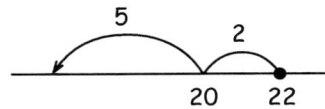

Other examples 43 – 8   61 – 6   72 – 5   91 – 7   73 – 4   82 – 3

Subtracting tens then units, not crossing a multiple of 10

Example 47 – 16   Where are you after moving back 10?

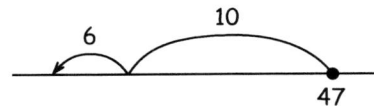

Example 98 – 43   Where are you after moving back 40?

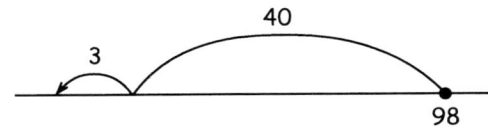

Other examples 74 – 43   89 – 45   78 – 23   99 – 62   58 – 34   49 – 11

Subtracting 9s by subtracting 10s then adjusting

Examples 24 – 9   78 – 49

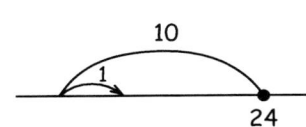 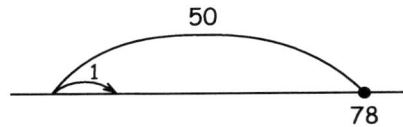

Where does the first jump take you?

Other examples 56 – 9   78 – 29   50 – 19   76 – 39   85 – 49   74 – 59

Subtracting the nearest multiple of 10 then adjusting

Examples 88 – 37   92 – 28

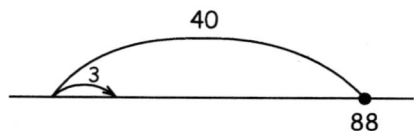

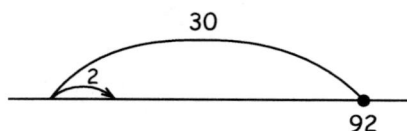

Other examples 59 – 28   54 – 37   84 – 68   45 – 28   81 – 66   77 – 48

Subtracting tens then units, crossing a multiple of 10

Example 62 – 26

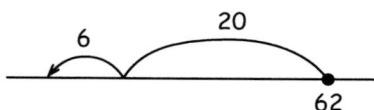

Other examples 73 – 34   41 – 27   65 – 38   72 – 33   55 – 17   82 – 37

Adding on to find the difference

Examples 88 – 37   92 – 28

Other examples 33 – 18   46 – 28   81 – 37   73 – 36   45 – 26   55 – 17

Subtracting from 100 (change from £1, £5, etc.)

For pupils who find this difficult, the 'adding on' method is probably easiest. For example, 100 – 57 can be shown like this:

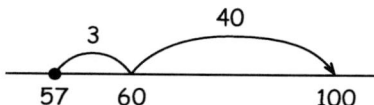

With practice, pupils may be able work out change from 100 like this:

100 – 57 is 'forty something': 10 – 7 is 3, so the result is 43.

Doubling

Doubling numbers between 1 and 50

Some doubles may be memorised already. Others can be worked out by doubling the tens digit first:

double 43 = double 40 + double 3

More difficult doubles require some mental addition:

double 36 = double 30 + double 6 = 60 + 12

Doubling numbers between 51 and 100

This involves knowing the doubles of 60, 70, etc. as well as mental addition.

For example, double 68 = double 60 + double 8 = 120 + 16 = 136

Multiplying by 4 and by 8

Examples $12 \times 4 = 12$ doubled, then doubled again

$13 \times 8 = 13 \times 2 \times 2 \times 2$

Halving

Halving even numbers to 100

Example   half of 58 = half of 50 + half of 8

or   half of 58 = half of 60 − half of 2

Finding a quarter by halving and halving again

Example   quarter of 56 = 56 halved (28) halved again (14)

Halving odd numbers to 100

Example   half of 27 = half of 26 + half of 1 = $13 + \frac{1}{2}$

or   half of 27 = half of 20 + half of 7 = $10 + 3\frac{1}{2}$

Multiplying

Unit 6 section A provides a suitable starting point. The methods described below can be looked at in short sessions spaced out over time.

Multiplying by 5 by multiplying by 10 and halving

Example $5 \times 42 = 420 \div 2 = 210$

Multiplying by 11 by multiplying by 10 and adjusting

Example $14 \times 11 = 14 \times 10 + 14 = 140 + 14 = 154$

Multiplying by 9 by multiplying by 10 and adjusting

Example $14 \times 9 = 14 \times 10 - 14 = 140 - 14 = 126$

Multiplying one-digit × two-digit

Mentally it is easier to multiply the tens first.
Example $5 \times 34 = (5 \times 30) + (5 \times 4) = 150 + 20 = 170$

Multiplying one-digit × two-digit using subtraction

To multiply by, e.g., 39, multiply by 40 first, then adjust.
Example $6 \times 39 = (6 \times 40) - 6 = 240 - 6 = 234$

Oral and mental starters

An oral and mental starter can be used for a number of purposes.

- It can **introduce the main topic**.
- It can also be an effective way of **revising skills that are needed for the main topic**.
- Alternatively it can be used to **'keep alive' skills learned earlier that are unrelated to the main lesson**.

Unit 2 'Number bites' and the oral questions in units 8, 17, 28 and 38 can be adapted into revision starters.

Many of the teacher-led activities in this guide can be used as oral and mental starters, for example 'Check it out' at the beginning of unit 19 'Brackets'.

Starters can also be based on questions in the pupils' book. For example, for questions C1 and C2 on page 78 pupils could come to the front and draw a rectangle of given area on a centimetre square grid on an OHP.

Starters often lead to useful discussion. All pupils should be encouraged to participate in them.

Starter formats

The formats described below have been found very effective and can be adapted to different topics.

Show me

You ask a question and pupils write their answer on a small whiteboard, or they choose a card from a prepared set and show it to you, giving you instant feedback on everyone in the class. Take advantage of opportunities to discuss ideas: for example, 'Show me a factor of 6' gives you a chance to establish that 1 is included in the factors of 6.

'Show me boards' work well with other types of starter, such as *True or false?* and *Odd one out*.

Today's number is …

Write a number on the board and put a ring round it. It could be a whole number, decimal or fraction. Pupils make up calculations with that number as the result. Calculations can be restricted to a particular type. (See also page 36 of this guide.)

Spider diagram

You write a whole number, fraction, decimal, percentage, word or algebraic expression in a circle on the board (the 'body'). The ends of the 'legs' can be completed in a variety of ways.

This is how it can be used to revise fractions of a number. Alternatively '$\frac{1}{2}$ of', for example, can be written in the body with quantities on the legs.

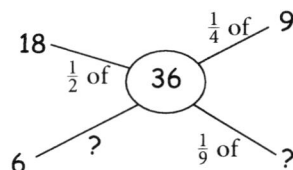

$$18 \quad \frac{1}{4} \text{ of } 9$$
$$\frac{1}{2} \text{ of} \quad \boxed{36}$$
$$6 \quad ? \quad \frac{1}{9} \text{ of } ?$$

Array

Prepare a grid or list of numbers beforehand. This could, for example, contain nine two-digit numbers: point to one of them and ask pupils to say what number added to it makes 100.

Alternatively, with appropriate grids, ask questions such as

- Give me two numbers in the grid that add to make $^-1$.
- Give me a square number.

Display

Display a diagram, graph, calculation, etc. (an OHP is often ideal for this); ask questions about it or ask pupil to make statements about it.

Odd one out

Write a list of numbers, times, words, pictures of shapes etc. Pupils have to identify which is the odd one out. For example,

- 1.3 m, 1.30 m, 1.03 m, 130 cm
- 9, 4, 25, 36, 12   (All except 12 are square numbers.)

Counting stick

Available through the usual equipment suppliers, this is marked in equal intervals, usually in two alternating contrasting colours. It is useful for counting-on activities. Give one end of the stick a number and tell the pupils the value of one interval; then ask them to count each number aloud as you point to the graduations in turn.

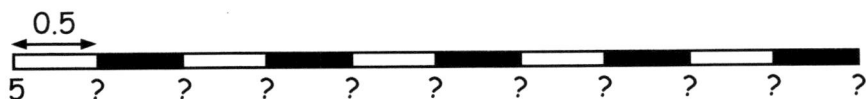

Alternatively give the values of two key points (you can Blu-tack labels on) then point to graduations at random, each time asking for the value.

True or false

Pupils have to decide if statements are true or false. Examples are:

• 9 is a factor of 3.

• This is the net of a cube (with an appropriate diagram).

Loop cards

These can be bought from suppliers, though you can easily make up sets of your own. Each card has a question on it together with the answer to a question that is on another of the cards; the complete set forms a loop.

Give out the cards. One pupil reads out their question; the pupil with the answer responds and then reads out their own question, and so on. To avoid leaving anybody out, make sure that there are at least as many cards as pupils. If there are more, then some pupils can have two cards.

Bingo

There are many topics where using a 'bingo' card to tick off answers works well. 'Tables bingo' in unit 6 'Multiplication' (page 56 of this guide) is one example. Getting pupils to write their own numbers from within a given range saves preparation and can often lead to useful discussion about what were the best numbers to have chosen.

Around the world

One pupil is chosen to start who then stands behind another pupil. You ask the two pupils a question and whoever answers first then stands behind a different pupil and so on. A number of topics can be covered by this method in one round.

Matching

Pupils have to match together pairs or groups of items that are the same, for example, pairing together divisions that have the same remainder. The items can be displayed. Alternatively a set of cards can be prepared and each pupil in the class given one; pupils then have to find who has the card(s) matching theirs.

Target number

Give the pupils four digits, which could be produced at random by calculator or a dice numbered 0 to 9. Pupils then have to write a calculation which is as close as possible to a given target. This is a useful way of practising brackets and order of operations.

Ordering

Present pupils with a set of items which they have to put in order, for example a set of rectangles with given dimensions to put first in order of size of perimeter and then in order of size of area. These can be on large cards held up at the front by a group of pupils who have to arrange themselves in the right order.

Hot seat

Choose three pupils to sit at the front in the hot seats. The rest of the class formulate questions on a particular topic to ask the panel. The activity could make use of the oral question stimuli on pages 53, 104–105, 169 and 227 of the pupils' book; pupils are likely to devise better questions if you have previously asked questions about the stimulus page (see, respectively, pages 66–67, 99, 141 and 180 of this guide for specimen questions).

Topics for starters

Below, arranged by broad topic area, are some suggestions for other ways these starter formats can be used.

Addition and subtraction

Today's number is … Pupils have to make addition and subtraction sums with a given answer. To avoid triviality you could insist that two-digit numbers are used.

Multiplication and division

Spider diagram Pupils multiply the body number by numbers on the legs. Alternatively write 'double' in the body with different numbers on the legs. Double could then be changed to 'double and double again' or 'halve'.

Odd one out Use a list of numbers that are all, except one, in a particular multiplication table.

Array Pupils multiply or divide each number on a grid by 10, 100 or 1000.

Loop cards Questions ask pupils to double, halve, multiply by 10 … the previous answer.

Show me Ask for an estimate for the answer to a given calculation.

Decimals

Counting stick Label the ends of the stick with decimals. Point to graduations in between, asking for each value.

Ordering Use cards with decimals on them.

Target number Use the approach of 'One or two' in unit 29 (page 170 of the pupils' book) but change the target to 5, 10 or some other number.

Fractions

Odd one out Show fractions (half, quarters and eighths only at this stage) that are all equivalent except one.

Spider diagram Put a number such as 48 in the body and have '$\frac{1}{4}$ of', '$\frac{1}{3}$ of' etc. on the legs.

Show me Ask pupils to shade a particular fraction of a simple rectangular grid.

Money

Hot seat Pupils can ask questions about unit 17 'Oral questions: money 1' (pages 104–105 of the pupils' book) or a similar resource.

Negative numbers

Counting stick Label a graduation in the middle of the stick as zero and one to its left as, say, ⁻2. Point to other graduations and ask what each represents.

Ordering Use cards with numbers that include negatives.

Show me Give out a set of cards with, say, the numbers from ⁻15 to 15 on them; ask 'Who can show me a number less than ⁻5?' etc.

Number relationships

Show me '… factors of …', '… multiples of …' and '… a prime number between …' work well.

Show me Give the pupils cards with numbers of a particular sequence. Each pupil holds their number up when it is the next one in the sequence.

Ordering Give a small group a set of cards with numbers of a particular sequence. The group puts them in order. New pupils with blank whiteboards then come up and add numbers to the sequence.

Time

Hot seat Pupils use the clock faces in unit 2 'Number bites' T1 (p 21 of the pupils' book) as a basis for questions.

Spider diagram Put a time in the middle and on the legs have '10 minutes after' and so on.

Matching Show pupils a list of times in different formats and ask them to match those that are the same.

Ratio

Around the world Using the recipes on page 238 of the pupils' book ask questions such as 'How much self raising flour do I need to make 20 scones?'

Order of operations

Target number Pupils are given four digits with which to make a given target number using operations and perhaps brackets.

Rounding

Bingo Place value bingo is described on page 87 of this guide.

Counting stick Label the ends of the stick as, say, 90 and 100. Point to a graduation and ask pupils 'Which number am I pointing to? Which is the nearest ten?'

Array Read out a number, say 148, and ask pupils which number on the grid is that number to the nearest ten.

Around the world Ask, for example, 'Which number is the nearest ten to 348?'

Symmetry

Odd one out Display a set of shapes which all, but one, have a particular type of symmetry, for example just one line of reflection symmetry.

True or false Display a shape, for example a parallelogram, and make statements such as 'This shape has two lines of symmetry.'

Show me Either place a set of cubes on an OHP or use a square grid. Make a shape to which pupils have to add a square to give specified symmetry. This could be continued with the shape repeatedly having squares added to give, or maintain, a particular symmetry. Alternatively use examples like those in section E of unit 5 (p 38 of the pupils' book).

Shapes

True or false Make statements such as 'An equilateral triangle has only two sides the same length.'

Show me Each pupil has set of cards with shapes on them. Ask questions such as 'Show me a scalene triangle' or 'Show me a shape with no lines of symmetry'.

Display Show pupils a design of the kind shown here. Ask them to identify different types of triangle.

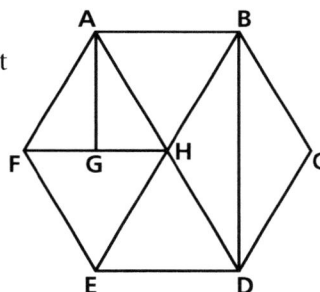

Angle

Estimating Display angles on an OHP for pupils to estimate.

Matching Display a set of labelled lines where pairs are perpendicular or parallel to each other. Ask pupils to identify a line which is parallel or perpendicular to a given line.

Array Display some labelled angles between 0° and 180°. Ask questions such as 'Which of the angles would make a straight line with 54°?'

Matching Pupils find pairs of angles that add up to 180°.

Display Show a diagram like the one above; pupils identify parallel and perpendicular lines.

Coordinates

Bingo See page 80 of this guide for a description of coordinate bingo.

Display Display a coordinate grid and play 'Four in a line' as described on page 80 of this guide.

Area and perimeter

Matching Show pupils a set of rectangles with given dimensions. Ask them to match pairs with the same area or perimeter.

Ordering Pupils order a set of rectangles or composite shapes in order of size of area or perimeter.

Metric units

Matching Use pairs of equivalent measurements such as 1.05 m and 105 cm.

Data handling

Array Show a grid depicting spinners with different numbers of sectors and different colourings. Ask pupils to show you a spinner with a probability of $\frac{1}{4}$ of red and so on.

Display Display a bar chart or pie chart and ask pupils questions about the information that can be obtained from it.

'Can you ...?' general assessments (sheets 3–6)

There are two of these, A and B, in the resource sheet master pack designed to help you identify, early in year 7, pupils with difficulties in some areas of basic number and to match number work to their needs. They are meant to be presented to pupils in a low key way 'just to give me some idea of what you can do' rather than as tests where something is at stake.

'Can you ...? A' is easier than 'Can you ...?' B. They are both to be done without a calculator.

The items are not grouped according to numerical operations because that might give too many clues when the aim is to find out whether the pupil can use the right one. So you may need to scan the completed test carefully to form a profile of the pupil's difficulties.

> 'Most of my students got full marks or just made careless mistakes – BUT the weaker students did show up.'

This information should help you plan how you will use the materials for low attainers in *Book 7T* (and *Book FN* if you are using it in year 7). It should help you start at the point where the pupil begins to have difficulties and prevent you from having to back-track further than necessary.

Ⓐ 'Can you ...?'

A1 (a) 8   (b) 14

A2 7

A3 (a) 4   (b) 4

A4

A5 (a) 13   (b) 204

A6 £2.20

A7 (a) $\frac{1}{2}$ (b) $\frac{1}{4}$ (c) $\frac{3}{4}$ (d) $\frac{1}{2}$ or $\frac{3}{6}$

A8 (a) $4 \times 2 = \mathbf{8}$ (b) $\mathbf{6} \times 2 = 12$
     (c) $5 \times 3 = \mathbf{15}$ (d) $4 \times \mathbf{5} = 20$
     (e) $10 \times 9 = \mathbf{90}$

A9 (a) 33   (b) 78   (c) 59

A10 £25

A11 (a) Two o'clock or 2:00
      (b) Twenty-five past two or 2:25
      (c) Half past 10 or 10:30
      (d) Quarter to twelve or 11:45

A12 (a) 5189   (b) 2060

A13 (a) 46   (b) 63

Ⓑ 'Can you ...?'

B1 83

B2 (a) 409   (b) 473

B3 24

B4

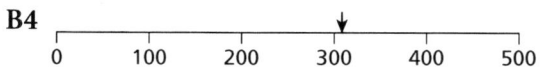

B5

×	4	5	6
2	8	**10**	**12**
3	12	**15**	18
4	**16**	**20**	**24**

B6 6 kg

B7 20 bars

B8 1×12, 2×6, 3×4, 4×3, 6×2 or 12×1

B9 The pupil's shaded fractions

B10 (a) 5   (b) 3   (c) 2

B11 (a) 3   (b) 2

B12 4

① **First bites**

This is a collection of activities suitable for use in the first one or two weeks of year 7. Their purposes are

- to give pupils of all abilities an enjoyable and confident start to mathematics in the secondary school
- to give you a chance to get to know the pupils and how they work
- to help establish classroom routines and ways of working (whole class, group, individual)
- to give opportunities for homework

They do not not need a high level of number skill, so should be widely accessible.

There may not be time to do all the First bites activities at the beginning of year 7. Some may be left for later, perhaps as starting points for related units of work.

Essential	**Optional**
Sheets 45 and 46	Sheets 47, 48, 49 (blank grids) and 58
Up to 15 dice for a class of 30	Square and triangular dotty paper (sheets 1 and 2),
Sharp pencils, pairs of compasses, rulers,	3 by 3 pinboards, rubber bands, tracing paper,
coloured pencils, board compasses	OHP transparency of square dotty paper
Angle measurers	

𝔸 **Spot the mistake** (p 4)

> Sheets 45 and 46

These two resource sheets give pupils of all abilities an opportunity, in a light-hearted context, to spot some mathematical errors – and some non-mathematical ones! It is more fun for pupils to work in pairs, rather than on their own.

𝔹 **Four digits** (p 4)

This activity will tell you something about pupils' knowledge of number operations and symbols (for example, brackets) and their arithmetic skills. It also gives an opportunity for co-operative group work.

◊ Decide together on the four digits to be used. They do not have to be all different. 0 is not very helpful.

◊ You can allow free rein at first as far as the rules are concerned. Pupils will probably come up with some ground rules themselves, and then you can establish rules for everyone, for example:
  • All four digits must be used.
  • No digit can be repeated unless it occurs twice in the set.
  • Digits can be used in any order.
  • Any operations can be used. (Brackets may be needed and √ may be suggested by pupils.)
  • Digits can be combined to make two- and three-digit numbers.
  • Results must be whole numbers (for example, $43 \div (2 + 1) \neq 14$).

◊ You could start by asking for ways to make, for example, 10.

◊ Pupils could work in groups, each group making a collection. An element of competition could be introduced. Alternatively, groups could be given ranges of numbers (1–20, 21–40, …).

◊ It is necessary to record the completed numbers and the methods used to arrive at them, for example, a list:

1	11	21
2	12	22
3	13	etc.
4	14	
5	15	
6 $1 + 3 + 4 - 2$	16	
7	17	
8 $2 + 3 + 4 - 1$	18	
9	19 $12 + 3 + 4$	
10 $1 + 2 + 3 + 4$	20	

◊ In one school the results were recorded on a large chart and put on the wall. This was added to over the year. (It works particularly well if there is some reward for completing gaps – merits, credits, etc.)

◊ Calculators may be used if necessary, although many pupils should do well without them.

◊ Pupils may realise that results often come in pairs, for example:
$$23 - 14 = 9, \quad 23 + 14 = 37$$
then with some swapping around:
$$32 - 14 = 18, \quad 32 + 14 = 46, \quad \text{etc.}$$

Follow-up

Each pupil can choose their own set of four numbers. Some sets of numbers (for example, 6, 7, 8, 9) are more difficult than others and may lead to demotivation.

ℂ **4U + 1T** (p 4)

This will identify pupils who have difficulty with place value in two-digit numbers

◊ Start by asking for a two-digit number. (If 13, 26 or 39 are suggested, find a sneaky way of avoiding them!) Write the number on the board. Say that you are going to multiply the units digit by 4 and then add on the tens digit (so 37 becomes $4 \times 7 + 3$, giving 31). Write the result on the board.

Ask pupils to do the same to this new number. Write the result. Repeat a few times until all pupils have understood the rule for generating the next number. Do not go on too long or you will form a loop. It is best for pupils to discover the loop for themselves.

You may come across a single-digit number in the process. If not, you should introduce one and discuss how to deal with it.

Ask pupils each to choose a two-digit number of their own, use the rule to make a chain of numbers and watch what happens as their chain grows.

After a while, someone will notice they have come back to a number they had before. When this happens, discuss it with the class. Pupils should realise that once a loop is formed, no new numbers are generated, but they do not always find this obvious!

◊ Questions for investigation are:

• Will all numbers form loops?

• How long are the loops?

• How many different loops are there?

• Do any numbers go straight to themselves? (Yes: 13, 26 and 39!)

Extensions

There is nothing special about 4 as the multiplier: it just gives short chains. Pupils can investigate other multipliers. How many chains are there in each case? (The rule 2U + 1T produces a nice overall diagram with every number connected to one big loop, except for the multiples of 19.) Pupils could be challenged to work backwards. For example, if the rule is 2U + 1T, what numbers generate 15 as their next number?

𝔻 **Finding your way** (p 5)

This gives practice in using left and right and in reading a simple map.

◊ Before discussing the picture and the questions, you could get a pupil to face the rest of the class, and ask which side of the room is on his or her left. ('Simon says' is also good for practice.)

◊ You could use a plan of your school. Give pupils a list of instructions from the classroom to somewhere else and ask them to guess the destination. They can then make up instructions for one another.

You could ask pupils to shut their eyes and imagine where they are going as you give them instructions.

◊ Before doing questions D4 to D7 you could do some practical work using the layout of the classroom. The aisles can be roads and the desks houses.

Someone, designated to be a 'Robot', can be given instructions to move about the class. Initially you could be the robot and ask a pupil to give you instructions. Then pupils can take turns to work in pairs.

D8–D11 Similar work can be based on local maps.

𝔼 **Gridlock** (p 7)

This game gives you an opportunity to find out pupils' addition (and subtraction) skills. Pupils can also develop and explain strategies to win.

> Up to 15 dice for a class of 30
> Optional: sheets 47, 48 and 49 (blank grids)

◊ To help them understand the scoring system, pupils could complete the grids on sheet 47 and work out the scores. Some schools have used this sheet for homework.

'I almost didn't use sheet 47 but it turned out to be most useful.'

◊ Initially the class could play together, with you rolling the dice and calling the numbers. Then the game can be played in groups of two or more.

To play 'Gridlock'

Each pupil draws a square grid (start with 3 by 3 grids) and marks off the top left-hand section as shown.

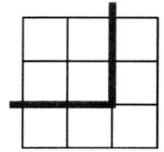

The caller rolls a dice and calls out the number. Each pupil writes the number in any empty square in the section shown shaded on the right.

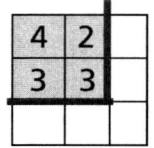

Repeat until each square in that section is filled.

Each number must be written in the grid before the next is called and a number can't be changed once it is written.

Each pupil adds up their numbers in the rows, columns and diagonal and writes the totals in the empty squares as shown on the right.

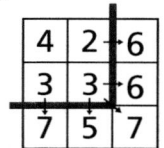

Each pupil adds up their points.

• Score 2 points for a total that appears twice.
• Score 3 points for a total that appears three times.
• Score 4 points for a total that appears four times ... and so on.

The grid above scores 4 points (6 and 7 both appear twice as a total).

After a number of rounds (decided by you), pupils add up their points and the one with most points is the winner.

◊ After playing on 3 by 3 grids, pupils could play the game on larger ones.

◊ Once pupils have played the game a few times, ask them to describe any strategies they use in placing the numbers on their grids. For example, if a number is rolled twice it is better to place the numbers diagonally,

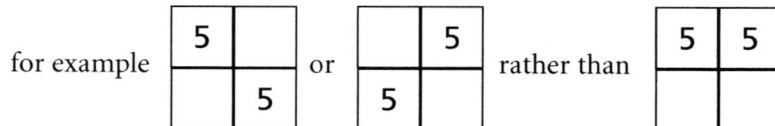

for example

5	
	5

or

	5
5	

rather than

5	5

◊ Now you can alter the rules as follows. First, the numbers called out are written at the side of the grid. When all numbers have been called, they are then placed in the grid. Pupils can think about how to get the maximum possible score with a particular set of numbers.

◊ One variation is for the winner to be the person with the fewest points. Pupils can discuss how their winning strategies change in this case.

Another variation is to use two dice to generate larger numbers.

Follow-up

In E1 to E11, the later questions are more difficult.

Remind pupils that they can only use numbers on an ordinary dice (1 to 6) to solve these problems.

E1 In (b), emphasise that their problems should be solvable without any guesswork or mind reading! Encourage more able pupils to make up problems that give the minimum necessary information.

This could be set as a homework task.

◊ You could ask pupils how changing the system of scoring points would affect the game, for example:

- score 2 points for a total that appears twice
- score 4 points for a total that appears three times
- score 6 points for a total that appears four times ... and so on

◊ A different version of the game is for pupils to cross out any totals that repeat and to add the remaining totals to give their score for that round. The winner could be the person with the most or fewest points.

𝔽 **Patterns from a hexagon** (p 9)

This work is to help pupils develop skills with compasses and rulers that are needed later to construct triangles and angles. Pupils also analyse patterns and make decisions about how to construct them.

At the start of the year many pupils have coloured pencils and geometry sets, so capitalise on this.

> Sharp pencils, pairs of compasses, rulers, coloured pencils, board compasses

◊ Many pupils find it difficult to draw a circle with a pair of compasses. They may need to draw circles and simple patterns before they feel confident enough to try the more difficult patterns.

Many pupils will find it helpful to see a demonstration of how to draw a regular hexagon. They must be able to draw a regular hexagon in order to draw the patterns on page 10.

Common problems include:

- not realising that the point of the compasses is moved to the point where the last arc crosses the circumference (and not at the end of the arc) for subsequent arcs to be drawn
- not realising that, to draw the hexagon, you join points where the arcs cross the circumference (and not the ends of the arcs)

Although pupils may have drawn them before, it may be helpful to demonstrate on the board or OHP how to draw the seven-circle or petal designs shown below.

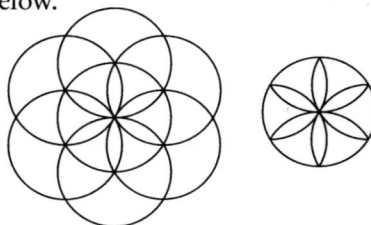

◊ This work provides good material for wall displays. In one school, the hexagon designs were used to make mobiles.

◊ You may want pupils to leave construction lines so you can check their methods.

◊ The construction of the patterns on page 10 offers more of a challenge, and the construction gets more involved further down the page.

The last two designs can be drawn as follows:

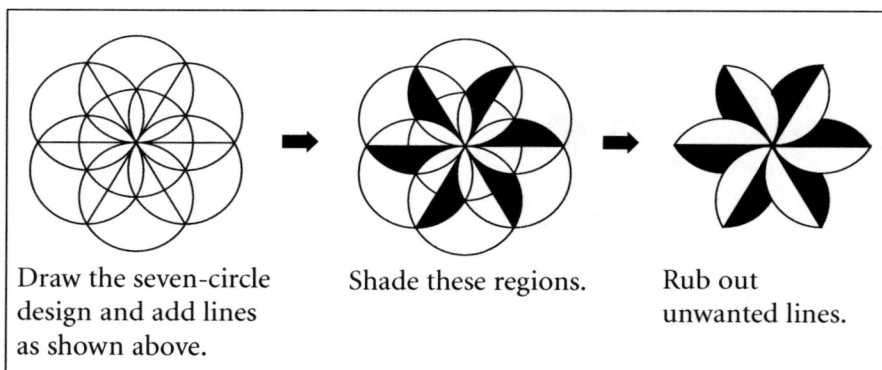

Draw the seven-circle design and add lines as shown above.

Shade these regions.

Rub out unwanted lines.

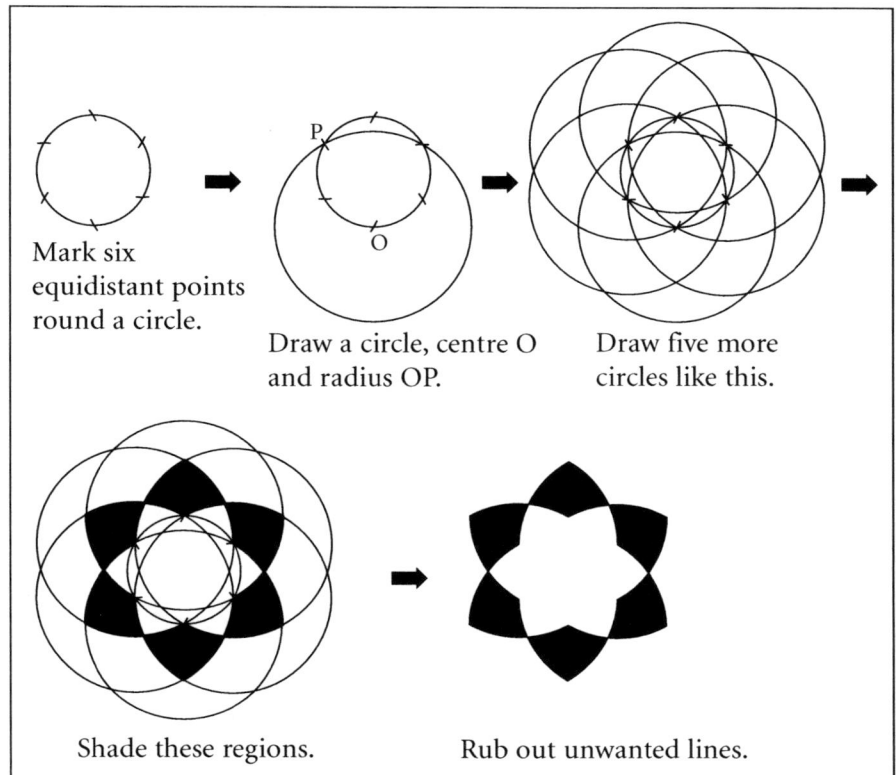

Mark six
equidistant points
round a circle.

Draw a circle, centre O
and radius OP.

Draw five more
circles like this.

Shade these regions.

Rub out unwanted lines.

Follow-up

If they know about symmetry, pupils could try to draw a pattern with
0 lines of symmetry, 1 line of symmetry etc.

Ⓖ **Shapes on a dotty square** (p 12)

Pupils create shapes and use mathematical language to describe them.
They can also decide on their own lines of investigation.

> Optional: square and triangular dotty paper (sheets 1 and 2), sheet 58,
> 3 by 3 pinboards, rubber bands, tracing paper, OHP transparency of
> square dotty paper

◊ Square dotty paper can be used or pupils can draw grids of dots on square
paper. Sheet 58 has the grids already ruled off.

◊ Establish rules for drawing shapes on the pinboard/grid.

• Only the 9 pins/dots can be used.

• All corners must be at a pin/dot.

• The types of shape shown in the pupil's book are disallowed (ones with
'crossovers' or 'sticking out' lines).

*'For some pupils,
drawing out the grid
was the hardest part.'*

◊ Ask pupils to draw a few different shapes following the rules above.

There is likely to be some discussion on the possible meanings of 'different' and 'same' here. 'Different' is usually taken to mean non-congruent. Tracing paper helps pupils identify shapes that are the same.

Look at some of their shapes together.

- What properties have they got?
  (Number of sides, angles, symmetry, parallel sides, area, …)
- Do pupils know names for any of the shapes?
  (Triangle, quadrilateral, rectangle, parallelogram, hexagon, …)

◊ There are various ways to structure this activity. Some trial schools generated a collection of questions from which pupils chose. For example:

> What shapes have the most sides?
> How many different squares?
> How many different triangles?
> How many shapes have reflection symmetry?
> How many shapes have rotation symmetry?
> What different areas can you make?
> How many ways can you put the same triangle on the grid?
> What shapes can be made with 1, 2, 3, … right angles?

Pupils can choose a question (or pose one of their own) and write up their solution. These could then be displayed.

For example, the 23 different polygons with reflection symmetry are:

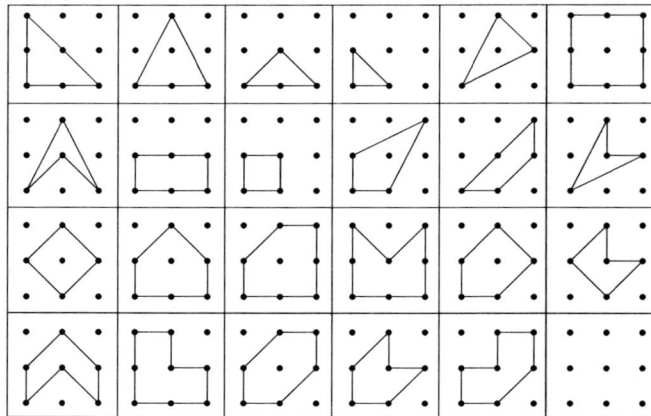

◊ An alternative structure is to begin by asking how many different triangles can be found. The 8 different triangles are:

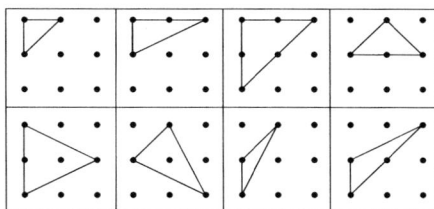

Now discuss the properties of the triangles. For example:

    Which has the greatest area?
    Which has a right angle?
    Which has reflection symmetry?
    Which are isosceles?

Pupils can now consider the different quadrilaterals that can be found and their properties. The 16 different quadrilaterals are:

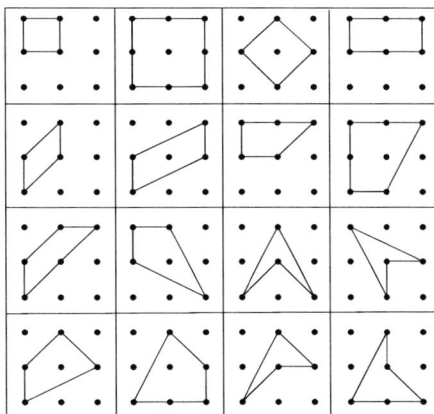

'I put the class into teams, following a lesson and a homework, to find all possible polygons, to collate results and present.'

Pentagons, hexagons and heptagons can be considered but the number of different shapes may be rather daunting!

The numbers of different polygons of each type are:

Number of sides	Number of different polygons
3	8
4	16
5	23
6	22
7	5

This gives 74 different polygons.

◊ One extension is to consider polygons on a different grid. Suggestions appear in the pupils' book.

The hexagonal 7-pin grid yields 19 different polygons which is a manageable number for most pupils to find. The polygons are:

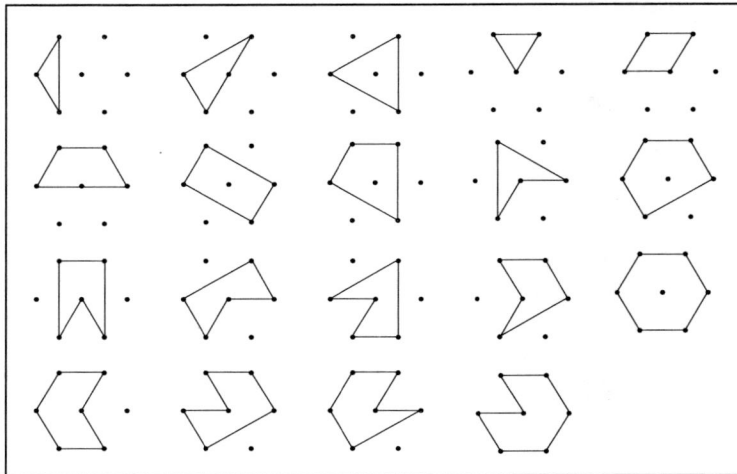

Ⓐ Spot the mistake (p 4)

1 Off to Benidorm in June

Fish tank framework is an impossible object.

Table left-hand rear leg is longer than the others.

Vase on the table is an impossible object.

Sofa is an impossible object.

Socket on left-hand wall has holes in wrong orientation.

Mirror reflection is incorrect; 'TAXI' and the clock face are the wrong way round.

Front door has handle and hinges on the same side.

The '33' above the front door is the wrong way round.

Vacuum cleaner plug has only one pin.

Vacuum cleaner hose has an extra hose tangled in it.

Triangles on shelf: right-hand one is an impossible object.

View of window blind is impossible.

2 Taxi to the airport

Clock has a back-to-front 3, and 8 where it should be 9.

P (parking) sign is on a road with double yellow lines.

Stop sign on road – S is wrong way round.

The word STOP (and the road marking) is on the wrong side of the road.

Bicycle has no front wheel.

Airport sign says 5 cm (centimetres).

Traffic lights have a right turn arrow to a no-entry street.

Rollerblader has one ice skate on.

Right-hand no-entry sign is pointing upwards.

Pedestrian crossing markings on the road should be rectangles.

Pedestrian crossing beacon is the wrong shape.

Street lamp on top of the no-entry sign is facing up.

A vegetarian butcher would not do much business!

Low bridge sign says min(imum) and should say max(imum).

Taxi is going to the airport, so should have turned left.

('Tax to rise 150%' is not necessarily wrong – it could!)

3 At the airport

A plane is flying upside down.

The plane taking off has no tail wings.

Passengers are walking along the wing of the waiting plane.

Waiting plane has RAF insignia on tail.

Waiting plane has a ski instead of a wheel.

Wind socks are blowing in opposite directions.

Tannoy message says 'train' instead of 'plane'.

Tannoy message contradicts time on clock.

Christmas tree contradicts Easter eggs sign.

Tax free sign: you cannot save 200%.

Suitcases are ticketed to Rome, and flight is to Benidorm.

Suitcase on weight machine has a square wheel.

Luggage weight sign says WAIT, not WEIGHT.

Luggage weight is in g(rams) and should be kg (kilograms).

The 'All departures' sign points to a no entry corridor.

You cannot 'Ski the Pyramids'.

You cannot 'Ice skate the Amazon'.

4 The hotel reception

Clock reads 14:60, and the minutes must be less than 60.

Sign above toilet doors says 'Welcome to BeMidorN.'

Calendar on reception desk says 31 June (only 30 days in June).

A Christmas tree in June is rare!

Toilet door pictures are the wrong way round.

Right-hand toilet door has handle and hinges on the same side.

Change sign – 100 pts should be 1000 pts, and the peseta has of course been replaced by the euro.

Double rooms are cheaper than single rooms.

Atlantic views should read Mediterranean views in Benidorm.

Left-hand rear leg on table is longer than other three.

Tickets on luggage have changed since the airport to Roma and Home.

Plant doesn't sit in its flower pot.

Lift is on ground floor (the lowest in the list above the door) but shows on the list (and the call buttons) as going further down.

List of floors above lift door is missing floor 4.

Sign in lift mirror – S is wrong in reflection.

Sign in lift has a weight limit that is silly. (The vase on the table is not an impossible object!)

5 By the pool

You don't get whales in the Mediterranean.

The flag at the top of the boat's mast should be blowing forwards.

Speed boat and water skier are not connected.

Plane is flying backwards if it is pulling the banner.

Banner should read 24 hours not 26.

Weather vane NSEW are wrong.

If the time is 23:30 the sun would not be out.

Temperature of $^-26°C$ is a bit chilly for sunbathing.

Water level in man's jug should be horizontal.

Sun lounger nearest to front is missing a leg.

Gazebo on the left-hand corner of the balcony is an impossible object.

Diving board heights are in millimetres, and should be in metres.

Lower diving board is at a greater height (8) than the upper (4).

There appears no means of access to the lower diving board.

'Do not feed fish' sign is unlikely in a swimming pool.

Man fishing not possible (we hope) in a swimming pool.

Swimming pool depth signs are wrong.

Shark in swimming pool.

Stairs and railings up to balcony create an impossible object.

Sangria jugs of 75 l(itres) would be a bit big!

Children shown standing at the left-hand side of the pool where depth is 5 m(etres).

Depth shown as 10 cm where the diving boards are.

Ⓓ Finding your way (p 5)

D1 Right

D2 Right

D3 Left

D4 (a) 2nd left   (b) 1st right
      (c) 3rd right   (d) 5th left
      (e) 4th right

D5 (a) Aspen Lane   (b) Birch Grove
      (c) 4th left

D6 Turn right into Beech Avenue, turn right when you get to Wood Street, take the 2nd right, Sam's is 4th house on the left.

D7 Turn right into Poplar Walk, turn left at Wood Street, take the 3rd right, Nikki's is 3rd house on the right.

D8 (a) Back Lane   (b) Downend Road
   (c) Left   (d) Right
   (e) Right   (f) Left

D9 Go along Windy Lane, take the 3rd turning on the left, the post office is on your left in Coronation Road.

D10 Turn left into Ashton Way, then turn right along Coronation Road, turn right, then 1st left, then 2nd left. Scott's is the house at the end of the road.

D11 Ashton Way and Mill Lane

E **Gridlock** (p 7)

E1 (a)

2	**3**	5
3	**4**	7
5	7	6

Points scored: **4**

4	**1**	5
3	6	9
7	7	**10**

Points scored: **2**

1	6	**5**	12
5	**3**	1	9
6	4	6	16
12	13	**12**	10

Points scored: **3**

(b) The pupil's problems

E2 Examples of grids that score 2 points are:

5	3	8
6	2	8
11	5	7

3	6	9
5	2	7
8	8	5

E3 Examples are:

(a)

6	6	12
4	3	7
10	9	9

(b)

4	6	10
6	3	9
10	9	7

E4 (a) 2 points   (b) 0 points

E5 (a)

4	**2**	6
2	**1**	3
6	3	5

(b)

3	**1**	4
2	**5**	7
5	6	8

(c)

5	**2**	**3**	10
4	6	**4**	14
1	**1**	**1**	3
10	9	8	12

E6 Examples are: 1, 2, 3 and 5; 2, 3, 4 and 6.

E7 The pupil's explanation

E8 The pupil's explanation

E9 (a)

6	**4**	10
4	**3**	7
10	7	9

(b)

1	**5**	6
4	**6**	10
5	11	7

E10 Examples are:

(a)

1	2	6	9
2	3	5	10
6	5	4	15
9	10	15	8

(b)

1	2	5	8
2	3	6	11
5	6	4	15
8	11	15	8

(c)

5	1	2	8
5	2	3	10
4	6	6	16
14	9	11	13

E11

1	5	5	11
1	3	4	8
6	5	6	17
8	13	15	10

Sheet 47

1

6	5	11
1	1	2
7	6	7

Points scored: **2**

2

3	4	7
6	5	11
9	9	8

Points scored: **2**

3

3	5	8
2	4	6
5	9	7

Points scored: **0**

4

3	4	7
4	3	7
7	7	6

Points scored: **4**

5

6	1	2	9
3	4	6	13
1	5	1	7
10	10	9	11

Points scored: **4**

6

5	3	5	13
3	1	4	8
2	6	1	9
10	10	10	7

Points scored: **3**

② Number bites

These activities are based on number skills that pupils will have met in key stage 2. *They are intended as a resource to be dipped into throughout the book and are not meant to be worked through as a single unit.*

A wide range of activities is included: written questions, oral work, puzzles and investigations. Many of the activities can be used in a variety of ways – as class activities, group work or individual tasks. It is hoped that the range of activities may suggest similar activities teachers could design themselves.

All the activities are expected to be done without a calculator.

The activities are mainly grouped into sections and the order of the activities represents a rough order of progression. Some later units rely on skills practised here; these may therefore provide useful revision practice for weaker pupils immediately prior to starting the units.

p 13	**P** Place value
p 14	**A** Addition and subtraction
p 16	**M** Multiplication
p 18	**W** Work-out – addition, subtraction and multiplication
p 19	**F** Fractions
p 21	**T** Time and money

Essential	Optional
Sheets 18, 19, 29 to 33, 36	Sheet 34
Ordinary dice and counters	Dice numbered 4 to 9 (one pair)
Sheets of rough or used paper	

ℙ Place value (p 13)

P1 Bigger wins (p 13)

Two sets of 0–9 cards per pair of pupils (made from sheet 19)

◊ You can make the rules of the 'Bigger wins' game clear by playing a demonstration game on the board with a pupil.

After playing for a while, most pupils realise they should put high numbers in the hundreds box and low numbers in the units box (and vice versa when putting cards in their opponent's boxes in the 'Nasty game'). Ask them to explain why.

P2 Mountains high (p 13)

The correct order is: Scafell Pike, Scafell, Helvellyn, Broad Crag, Skiddaw, Ill Crag, Helvellyn Low Man.

P3 Oceans deep (p 14)

The correct order is: Gulf of Mexico, Bering Sea, Mediterranean Sea, Arctic Ocean, Malay Sea, Caribbean Sea, Indian Ocean, Atlantic Ocean, Pacific Ocean.

𝔸 Addition and subtraction (p 14)

A1 Addition bingo (p 14)

> Two sets of cards numbered 4 to 9 cut from sheet 19 or two dice, each numbered 4 to 9

◊ Show pupils the numbered cards or tell them how the dice are numbered. Tell them that you are going to use these to choose two numbers and that they have to add these each time. Ask them to choose eight numbers to go on their card. After a few games, all the numbers they choose should be in the range 8 to 18! The winner is the first person who has correctly crossed off all the numbers on their card.

'Good idea and used a number of times. The class often asked for a game of "Bingo" at the end of the lesson as I usually gave a prize to the winner!'

A2 Today's number is ... (p 14)

Write a number on the board to be 'today's number'. (On later occasions this could be chosen by a pupil.) Pupils then make up calculations with that number as their result, or make true statements about the number using as great a variety of mathematical language as possible.

This activity can be extended beyond addition and subtraction to include other number skills.

A3 Minimal measuring (p 15)
(a) The 10 cm and 3 cm strips
(b) 4, 6, 8, 11, 13, 15
(c) (i) 9, 14, 16, 18   (ii) 19
(d) 2, 7, 12, 17
(e) 1, 2, 4, 8, 16 The longest length is 31 cm.

A4 Cover up (p 15)

Sheet 18

◊ Many schools found this to be a useful homework activity.

◊ Possible solutions to the puzzles are shown below.

A

13	11
15	
12	
16	17

(18 left side, 14 right side)

B

12		
11	14	17
18		
15		

(13 left, 16 right)

C

	12	
	16	14
18		
15	11	17

(13 left)

D

15	17	14
12	13	
		18
11	16	

M Multiplication (p 16)

M1 Today's number is … (p 16)

◊ To ensure that pupils do lots of multiplication, choose a number that has lots of factors!

M2 Grids (p 16)

◊ Many trial schools found this activity popular and useful. One school commented that it could be suggested to parents as a way of practising tables with their children.

M3 Multiplication dominoes (p 16)

Dominoes made from sheet 29 (one set per solo player or group)

The cards can also be used to play the competitive game of dominoes or pupils can cooperate to make a complete 'ring'.

M4 Multiplication pairs (p 16)

Cards made from sheets 30 and 31 and/or 32 and 33 (one set per group)

◊ There are two sets of cards:
  • 'Mults': easier tables (sheets 30 and 31)
  • 'Times': full range of tables (sheets 32 and 33)
You could copy each set on a different colour.

M5 Links, chains and loops (p 17)

Optional: sheet 34 (for recording solutions)

◊ This is a more demanding activity but pupils generally enjoy the challenge.

(a) (4) —28— (7)

(b) (5) —10— (2)

(c) (6) —24— (4)

(d) (3) —24— (8) —32— (4)

(e) (3) —18— (6) —42— (7) —35— (5)

(f) (8) 16 / 48 (2) —12— (6)

(g) (5) 30 / 40 (6) —48— (8)

(h) (9) —27— (3) 45 / 15 (5) —25— (5)

(i) (6) —36— (6) —42— (7) (6) —18— (3) —21— (7)

(j) (4) 20 (5) 40 / 35 (8) (7)

(k) (4) 20 / 12 (5) —15— (3)

(l) (3) 24 / 15 (8) —40— (5)

(m) (8) 32 / 72 (4) —36— (9)

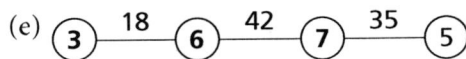

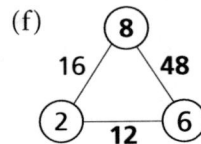

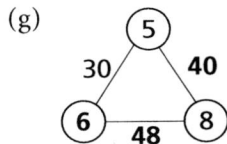

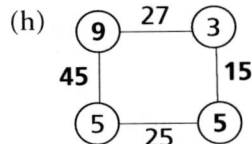

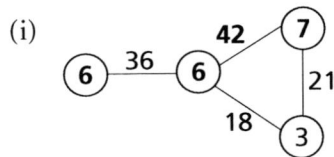

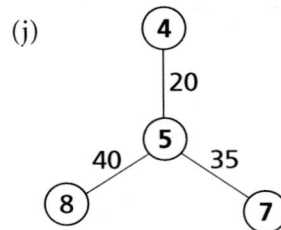

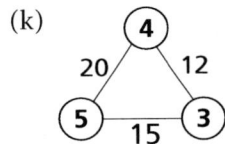

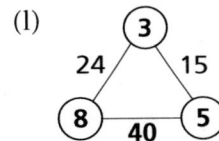

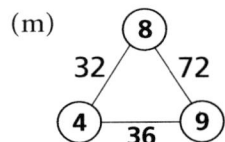

There are other solutions for (m)
but outside the usual range of the
multiplication tables.

𝕎 Work-out – addition, subtraction and multiplication (p 18)

W1 Three in a line (p 18)

> Two ordinary dice (per pair of pupils), eight counters of each of two
> colours (or bits of paper with initials)

◊ Once you have clarified the rules with pupils you should be able to spend
some time working one-to-one with weaker pupils. For example, one
teacher who trialled this activity set it up as a tournament for the whole
class to keep momentum going. She then played with the weakest student,
'who was quite happy to add but had to be encouraged to multiply'.

'We had quite a good
discussion about why
they couldn't get 13
or 14.'

W2 Pairs – an investigation (p 19)

◊ To get the largest result, you pair off the two largest numbers, then the
other two numbers.

F **Fractions** (p 19)

F1 Shade half, shade a quarter (p 19)

> Sheet 36 or squared paper

◊ Pupils need only do a few of these to get the point. More confident pupils can include diagonal lines.

F2 Naming parts (p 20)

> Sheets of paper, all the same size (rough or used sheets will do; each pupil needs two sheets)

◊ This activity can be demonstrated 'from the front' using a large sheet of paper.

Show the class how to fold one of their sheets into two equal parts and the other into four.

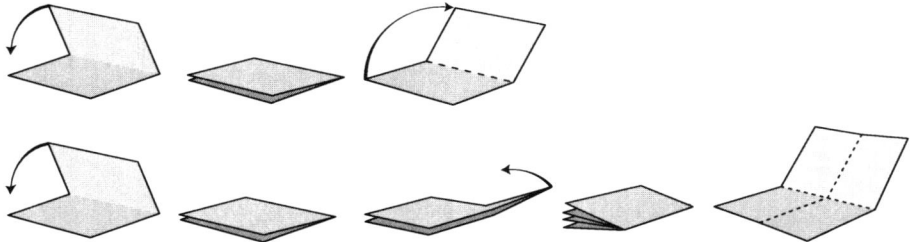

Open out the sheets. If pupils can't see the creases on your sheets, go over them with dotted lines.

Trace round one of the quarters with your finger (but don't use the word 'quarter' at this stage). Ask pupils to do the same and describe in different ways what they see. Pupils' descriptions have included

'1 out of 4'

'4 out of 4 is another name for one whole sheet.'

'4 quarters'

'One whole sheet divided by 4 is a quarter.'

'4 of the 1-out-of-4s is the same as 1 whole sheet.'

Ideas written on the board have included

$$\frac{1}{4} \qquad 4 \times \frac{1}{4} = 1 \qquad \frac{4}{4}$$

$$\frac{4}{4} = 1 \qquad 1 \div 4 = \frac{1}{4}$$

> *'Equivalent fractions arose naturally here ... The light of understanding dawned by having done something practical.'*

Now trace your finger around two quarters. Ask pupils to do the same and to talk and write about what they have done.

At some point, when pupils talk about a quarter, ask them 'a quarter of what?' and try to establish that fractions are fractions *of* something. Each pupil currently has two sheets the same size, but if they were different sizes it would no longer be true that two quarters of one sheet was the same as half the other.

◊ Identifying fractions is not just about counting parts. The parts have to be suitable ones and some pupils may have difficulties about this.

Fold a sheet like this in front of the class. Ask whether the two parts are halves.

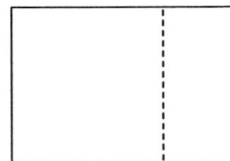

What about a sheet folded like this?

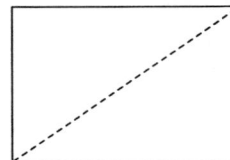

Fold a sheet corner to corner. If pupils say that the two parts are halves, challenge them to convince somebody (you or another pupil).

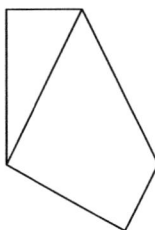

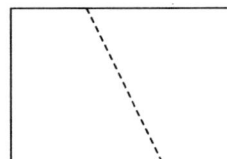

(a) Half, 1 out of 2, $\frac{1}{2}$

(b) 1 out of 4, a quarter, $\frac{1}{4}$

(c) $\frac{3}{4}$, 3 out of 4, three quarters

(d) A whole, two halves, $\frac{2}{2}$

(e) 2 out of 4, $\frac{2}{4}$, a half, $\frac{1}{2}$

(f) 4 out of 4, four quarters, a whole

(g) 2 out of 4, $\frac{2}{4}$, a half, $\frac{1}{2}$

(h) 3 out of 4, three quarters, $\frac{3}{4}$

(i) 1 out of 4, a quarter, $\frac{1}{4}$

(j) 3 out of 4, three quarters, $\frac{3}{4}$

(k) 2 out of 4, $\frac{2}{4}$, a half, $\frac{1}{2}$

(l) 2 out of 4, $\frac{2}{4}$, a half, $\frac{1}{2}$

T Time and money (p 21)

T1 Watch it (p 21)

The page is for teacher-led oral work on telling the time and carrying out simple calculations with time. It can be used to practise skills that pupils have already acquired and to introduce more demanding mental work on the topic.

◊ Not all pupils are fluent with both of the common time formats ('ten to six' and '5:50') so you should use both in your questioning.

A few pupils still have fundamental difficulty with time when they enter secondary school and this can be a source of acute embarrassment for them. Any questions you pose to them in a class session based on this page will need to be very simple. You will almost certainly need to organise one-to-one work with them to make significant progress.

These sample questions are roughly in order of increasing difficulty.

1 What time does watch C show? 4:30

2 Which watch shows five to six? I

3 Watch A is one hour slow. What is the real time? 4:00

4 Mr Jones takes some trousers in to be dry cleaned. Watch G shows the time. The trousers will be ready in 3 hours. When will they be ready? 5:45

5 You look at your watch and it looks like B. How long before it will look like D? 14h

6 Claire's friend promised to meet her at two o'clock. G shows the time and her friend still hasn't arrived. How late is she? $\frac{3}{4}$h

7 You phone your friend at a quarter to nine. Watch D shows the time when you finish the phone call. How long was the phone call? $\frac{1}{2}$h

8 Mrs Patel parks her car. Watch D shows the time. She pays for 45 minutes' parking. When must she get back to her car? 10:00

9 If the real time is ten past one, how fast or slow is watch F? 20 min slow

10 Sarah needs to catch a train that goes at 7:05. She looks at her watch and H is what she sees. How long is it before her train goes? 40 min

11 F shows the time when you put a cake in the oven. It takes 50 minutes. When is it ready? 1:40

12 H shows the time when you get on a bus. B shows the time when you get off. How long has your journey taken? 1 h 35 min

13 A TV programme starts at time I and ends at time D. Will a three-hour video tape be long enough to record it?

No, the programme is 3 h 20 min long

◊ You can devise questions that relate to the real school day, for example:

    1 If A shows the correct time, how long is it until the end of school?

    2 Do any of the watches show a time during today's maths lesson?

T2 Sweet tooth (p 22)

  1 (a) 92p   (b) 75p   (c) 84p   (d) 47p

      (e) 58p   (f) 87p   (g) 50p   (h) 47p

  2 (a) £4.63   (b) £4.84   (c) £4.53   (d) £4.67

      (e) £3.94   (f) £4.00

  3 A Freshers

  4 A Fudge Finger and a Freshers

  5 An Apple Crunch and a Fizzo

 *6 A Fudge Finger, a Tri Bar and a Fruit Chew

③ **Written addition and subtraction** 7S/3

Essential	Optional
Sheet 19	Coins (plastic or real)
	Dienes base ten equipment
	Sheets 24 and 25
	Dice
Practice booklet pages 3 and 4	

Ⓐ **Written addition practice** (p 23)

The 'Total' game involves addition only.

You may wish pupils to do the addition practice in questions A1 and A2 before playing 'Total'.

Total (p 23)

> A set of cards numbered 0 to 9 (sheet 19) or a dice (for each group)

◊   The game can, if necessary, be simplified by using fewer cards (for example, 0 to 5) or smaller grids.

◊   After a few games discuss any strategies pupils use in positioning the digits.

◊   If pupils are using dice instead of cards, discuss ways in which they would change their strategies if they used a different dice.
For example, using a 1 to 6 dice, if 5 was the first digit rolled where would they put it? If they used a 1 to 8 dice would they put a 5 in the same position in their grid?

◊   At the end of a game ask the pupils in a group to see if they can
- rearrange the digits they used to find a higher total
- all find a different way of getting the same total

◊ Pupils could play the game again but change the rules in some way. For example, what if the winner is the player
  • with the lowest score
  • whose score is nearest an agreed target number

◊ Ask pupils to make 1000 (or 100 at first) by choosing any of the cards. After some time discuss any strategies they used to find a solution. In particular discuss why the digits in the units column must add up to 10, and the digits in others must sum to 9.

(There are 192 different solutions for 1000, so do not expect them to find them all!)

B Written subtraction practice (p 24)

Those who have difficulty with written subtraction are likely to need much individual help.

> Optional: Dienes base ten equipment or 10p and 1p coins,
> a set of place value cards (such as those on sheet 25)

◊ The term 'borrowing' is very misleading. It gives the the impression of an unfinished transaction. A more appropriate term is 'exchanging': we exchange a ten for 10 units or a hundred for 10 tens.

◊ Point out that it is a good idea to check the answer to a subtraction by adding. For example, the calculation on page 24 can be checked by showing that $128 + 217 = 345$.

◊ Coins can be used to help understanding in two-digit subtraction.

For example, $35 - 17$ can be tackled by first showing that 3 tens and 5 units can be exchanged for 2 tens and 15 units.

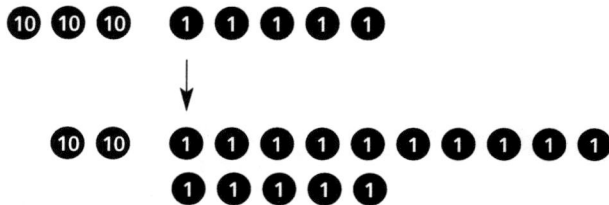

Now subtract 17 to leave 18.

C Money (p 25)

D Investigations (p 26)

Palindromes (p 26)

This investigation consolidates addition.

Optional: a hundred square (sheet 24) and coloured pencils

◊ After the class has worked together through a few starting numbers, pupils can choose some starting numbers of their own.

◊ Don't choose 89 or 98 as one of your early starting numbers!

◊ Pupils may notice that, for example,
  • multiples of 11 are already palindromes
  • a number whose digits add to 9 gives 99 after one loop
  • a number whose digits add to 10 (apart from 55) gives 121 after two loops
  • a number whose digits add to 11 gives 121 after one loop
  • a number with each digit less than 5 gives a palindrome after one loop

◊ Results can be recorded on a hundred square using colours to show how many loops are needed. Discuss the patterns as they start to appear. Why, for example, is there symmetry about a diagonal line? (The patterns are more obvious if the square is numbered 0–99 instead of 1–100.)

For single-digit numbers, you could say that 8 is already a palindrome. Or you could count 8 as 08, in which case 08 + 80 = 88, etc.

In the solutions below single digits are treated in the latter way.

1 to 100 square

A palindrome already	1	2	3	4	5	6	7	8	9	10
One loop	11	12	13	14	15	16	17	18	19	20
	21	22	23	24	25	26	27	28	29	30
Two loops	31	32	33	34	35	36	37	38	39	40
Three loops	41	42	43	44	45	46	47	48	49	50
	51	52	53	54	55	56	57	58	59	60
Four loops	61	62	63	64	65	66	67	68	69	70
Six loops	71	72	73	74	75	76	77	78	79	80
Twenty five loops	81	82	83	84	85	86	87	88	89	90
	91	92	93	94	95	96	97	98	99	100

0 to 99 square

0	1	2	3	4	5	6	7	8	9
10	11	12	13	14	15	16	17	18	19
20	21	22	23	24	25	26	27	28	29
30	31	32	33	34	35	36	37	38	39
40	41	42	43	44	45	46	47	48	49
50	51	52	53	54	55	56	57	58	59
60	61	62	63	64	65	66	67	68	69
70	71	72	73	74	75	76	77	78	79
80	81	82	83	84	85	86	87	88	89
90	91	92	93	94	95	96	97	98	99

Number magic (p 27)

◊ All pupils can try the investigation on two-digit numbers.

Extending to three or four digits may well prove too much for some pupils unless they are able to use a calculator.

Two-digit numbers

Pupils may notice that:
- The first answer is always in the 9 times table.
- The final answer is always 99 (treating 9 as 09).

Three-digit numbers

The final answer is always 1089 (treating 99 as 099).

Four-digit numbers

The final answer depends on the middle pair of digits. If the 2nd digit is greater than the 3rd, the result is 10890. If the 2nd and 3rd digits are equal, the result is 10989. If the 2nd digit is less than the third, the result is 9999.

𝔸 **Written addition practice** (p 23)

A1 (a) 195 (b) 319 (c) 372
   (d) 680 (e) 684

A2 (a) 432 (b) 332 (c) 282
   (d) 781 (e) 1020

𝔹 **Written subtraction practice** (p 24)

B1 (a) 254 (b) 347 (c) 356
   (d) 392 (e) 182

B2 (a) 614 (b) 237 (c) 292
   (d) 164 (e) 515

B3 (a) 466 (b) 588 (c) 475
   (d) 688 (e) 579

B4 (a) 423 (b) 417 (c) 105
   (d) 244 (e) 188

B5 (a) 173 (b) 355 (c) 859
   (d) 479 (e) 651

B6 (a) 152 (b) 346 (c) 468
   (d) 249 (e) 156

ℂ **Money** (p 25)

C1 (a) £2.91 (b) £2.02
   (c) £1.71 (d) £3.19

C2 (a) £1.42 (b) £1.29
   (c) £3.25 (d) £1.39

C3 (a) £2.69 (b) £8.67
   (c) £3.47 (d) £2.48

C4 (a) £1.60 (b) £4.29 (c) £2.28

C5 £0.35 or 35p

C6 £1.95

C7 £1.57

*C8 Coffee and a cheese ploughman's is the most likely lunch. Other combinations are possible, for example 3 lemonades, a ham sandwich and a tea cake.

What progress have you made? (p 27)

1 (a) 857 (b) 421 (c) 632 (d) 514
  (e) 316 (f) 177 (g) 341 (h) 177
  (i) 426

2 (a) £3.22   (b) £4.91

     (c) £2.87   (d) £1.44

Practice booklet

Sections A and B (p 3)

1 (a) The highest total is 573.
The possible sums are

541	532	542	531
+ 32	+ 41	+ 31	+ 42
573	573	573	573

     (b) The highest total is 1056, for example,

980	986	970	976
+ 76	+ 70	+ 86	+ 80
1056	1056	1056	1056

2 (a) 953   (b) 413   (c) 978

3 The highest total is 1804.

4 (a) 4   (b) 4   (c) 9

5 (a) 361 miles   (b) 21 miles

6 (a) 631   (b) 223   (c) 267

     (d) 145   (e) 459   (f) 507

     (g) 259   (h) 45

Section C (p 4)

1 £4.62

2 (a) £1.60   (b) £2.90

     (c) 35p or £0.35   (d) 19p or £0.19

3

+	29p	£3.22	**£1.16**
34p	**63p**	**£3.56**	£1.50
£1.37	**£1.66**	**£4.59**	**£2.53**

4 (a) £2.14   (b) £5.23   (c) £4.00

     (d) £4.15   (e) £8.29   (f) £8.43

5 (a) £3.15   (b) £3.82   (c) £6.17

     (d) £6.44   (e) £1.86

     (f) 45p or £0.45

6 (a) £8.20   (b) £4.54   (c) £4.13

     (d) £5.41   (e) £9.37   (f) £4.22

④ Test it!

Pupils collect measurements to test general statements.

p 28	**A** I don't believe it!	Planning a task Collecting data to test a statement Measuring
p 29	**B** Organising your results	Organising measurements
p 31	**C** Now it's your turn!	Testing a chosen statement

Essential

Metre sticks, tape measures (enough for at least one item per group)

Ⓐ I don't believe it! (p 28)

> Metre sticks, tape measures (at least one item per group)

◊ Organise pupils into small working groups. (Groups of four work well.)

After you have introduced the statement 'Everyone is six and a half feet tall', the groups can discuss the first set of questions.

It should become clear that

- six and a half feet tall means six and a half 'foot lengths' tall
- the statement can be tested by measuring or simply stepping off each pupil's foot length against their height

You could have a general discussion at this point comparing the groups' plans for testing the statement. Or the groups could move straight into carrying out their plans. Some pupils may need a lot of help with measuring.

Recording of data may be haphazard. This is taken up in section B, which some teachers have preferred to do before A.

Methods used by pupils include:

- making an outline of themselves on a roll of old wallpaper, then cutting out their footprints to test the statement
- Blu-tacking rulers to walls to make it easier to measure heights
- measuring out heights on the tape and then 'stepping off'

In work of this kind it is important to make a plan and to adapt it as necessary. In many cases pupils do not see the need for a plan, preferring instead to 'jump right in'. You may find examples to emphasise this point in the pupils' own work.

Encourage each group to compare findings with others.

Six and a half feet tall

◊ After this discussion it is worth raising the issue of whether the statement 'Everyone is six and a half feet tall' is true for a wider population. Pupils could investigate the statement by measuring younger or older people at home for homework.

> *'Pupils enjoyed this topic. However, there were several teething problems such as which groups they were in and who was responsible for what. When things settled down they produced some excellent work which was good for display. The groups also gave a presentation of their work.'*

B Organising your results (p 29)

◊ Some teachers have preferred to do this section before section A.

B1 Pupils should consider the problems of
- mixed units
- writing results in different orders

B2 You may need to help with 'approximately'.

C Now it's your turn! (p 31)

Metre sticks, tape measures (at least one item per group)

Pupils choose their own general statement to test.

◊ Each group must decide how to measure, for example, the 'length' of a person's head.

◊ A formal write-up is not necessarily expected at this stage. You could ask for a poster from each group or each pupil. (Later in the course there is a more specific focus on writing up results.)

B Organising your results (p 29)

B1 This method can be improved by presenting results in the same order and everyone using the same units.

B2 Ben is correct.

B3 (a) It may mean the height from the chin to the top of the head (when the mouth is closed).

(b) Tim's arm span might be 170 cm.
Gina's height might be 1.44 m.
Sue's height might be 1.59 m.
Sue's foot length might be 26 cm.
Ryan's height might be 1.5 m.
Lara's hand span might be 17 cm.

(c) Tim, Sue, Ryan (if he is 1.5 m tall) and Majid can go on the rides.

(d) Neena's arm span might be about the same length as Ajaz's, 153 cm.

(e) Depends on the pupil's own measurements

(f) About 6.4 to 7.3 times
(The height and head length have to be expressed in the same units before dividing.)

What progress have you made? (p 31)

1 Yes – it is true that heights are roughly twice the inside leg measurements.

⑤ Reflection symmetry

Pupils use folding, cutting and mirrors in a variety of practical activities. They have especially enjoyed the symmetry tiles game (section D) and it has proved very effective in consolidating the key ideas.

Rotation symmetry is not dealt with in this unit. However some pupils are likely to be aware of the idea and it is there as a 'distractor' in some questions. You may wish to discuss it where appropriate.

p 32	**A** Folding and cutting	Predicting a mirror image – checking by folding and cutting
p 33	**B** Using a mirror	
p 34	**C** Looking for reflection symmetry	
p 36	**D** Times and dates	Lines of symmetry in groups of digits
p 38	**E** Shading squares	Making symmetrical designs by shading squares

Essential	**Optional**
Mirrors	Tracing paper
Scissors	
Sheets 59 to 65	
Practice booklet pages 5 to 7	

𝔸 **Folding and cutting** (p 32)

Pupils predict a mirror image and check by folding and cutting (including reflecting in sloping lines).

Scissors, sheets 59, 60, 61, 62
Optional: mirrors, tracing paper

5 Reflection symmetry • **51**

> 'As an introduction, pupils drew the second half of a mask. They found it fun and the masks made a good display. They knew what to do in a non-mathematical context. This built confidence.'

◊ Pupils could use alternative methods (a mirror or tracing paper) but folding and cutting gives an immediate check that is very convincing. You may wish to allow a variety of methods in the classroom depending on pupils' prior attainment.

◊ Sheet 61 should reveal the common errors when reflecting in sloping lines. Counting dots outwards from the mirror line can be a useful approach.

B Using a mirror (p 33)

Pupils use a mirror to draw and check a mirror image (including reflecting in sloping lines).

Mirrors, sheet 63

◊ It may be worth leading the class through the stages in the photographs, stressing the meaning of the word 'image'.

C Looking for reflection symmetry (p 34)

Pupils identify lines of symmetry and use a mirror to check.

Mirrors

◊ It may be worth leading the class through the stages in the photographs.

C2 Some pupils may say that designs (d), (g), (j) and (l) have reflection symmetry because they recognise the rotation symmetry. It is a good opportunity to discuss this type of symmetry.

D Times and dates (p 36)

Pupils find lines of symmetry in groups of digits (times and dates) where two lines of symmetry are possible, and play a game.

Scissors, mirrors, sheet 64 (copied on card if possible), sheet 65 (one for each group of players)

◊ Many pupils find it harder to identify a horizontal line of symmetry than a vertical one. It may be worth leading a brief discussion after D1 has been completed.

◊ Some pupils may not be familiar with the method of writing dates used in D2 onwards. It may be beneficial to discuss this with pupils before they begin this work.

D4 Pupils may be uncomfortable about the answer 'none' for (c) as they may recognise the rotation symmetry. It is a good opportunity to discuss this type of symmetry.

◊ Emphasise that dates found in D6 and D7 have to be possible. For example, 83:38:83 is not a possible date!

Symmetry tiles game

Pupils can use this to consolidate their understanding of symmetry.

◊ The game should be self-correcting, with players protesting at an invalid move. Some care is needed in organising the groups: each group should have one pupil who is confident enough about symmetry to recognise invalid moves. Although the game can be played with four players, having only two or three makes it faster and more enjoyable.

◊ You may have to clarify one or two things: pupils can put their cards down on either side of the dotted line; they don't have to complete the symmetry at every move (though they could play that way).

◊ Watch for pupils who are still getting the symmetry wrong. Provide them with a mirror so they can see their mistakes.

◊ Boards 2 and 3 on sheet 65 are harder versions of the game.

E Shading squares (p 38)

Pupils make designs with one or more lines of symmetry by shading squares. They can try to devise strategies to ensure all the different ways of shading squares are included.

Optional: mirrors (to check results)

◊ Lower attaining pupils may feel more confident if they are told there are four ways for each problem in E1 and E2.

◊ These problems can be extended in a variety of ways. In one school, pupils suggested their own extensions and looked at shading different numbers of squares on these diagrams.

A Folding and cutting (p 32)

A1 to *****A5** The pupil's drawings and checks

B Using a mirror (p 33)

B1 to **B2** The pupil's symmetrical drawings

C Looking for reflection symmetry (p 34)

C1 (b) Yes (c) Yes (d) No (e) Yes
     (f) No (g) Yes (h) Yes

C2 (a) Yes (b) No (c) Yes (d) No
     (e) Yes (f) No (g) No (h) Yes
     (i) Yes (j) No (k) Yes (l) No
     (m) Yes (n) No (o) Yes (p) Yes

D Times and dates (p 36)

D1 (a) 01:18, 03:38, 11:18, 01:00, 13:13, 01:10 and 10:01 are symmetrical.

    (b) 01:10 and 10:01 have two lines of symmetry.

    (c) The pupil's three times with one line of symmetry

    (d) One time from 11:11 and 00:00

D2 Two lines of symmetry

D3 Yes, it is symmetrical.

D4 (a) None   (b) Two   (c) None
    (d) One

D5 (a) 04:02:33   none
    (b) 31:10:81   one
    (c) 08:11:80   two
    (d) 08:01:80   one

D6 The pupil's two dates with one line of symmetry

D7 The pupil's two dates with two lines of symmetry

E Shading squares (p 38)

E1 Four different ways

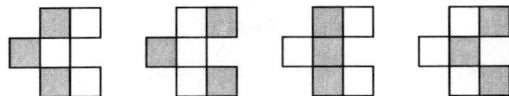

E2

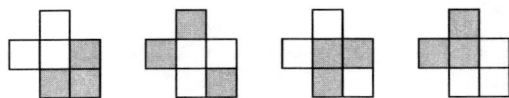

*****E3**

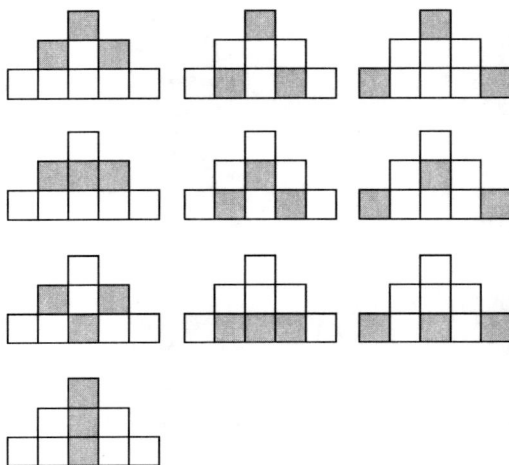

What progress have you made? (p 38)

1 (a)   (b)

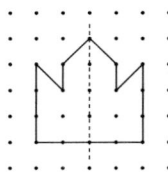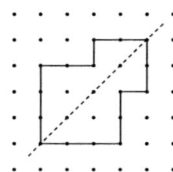

2

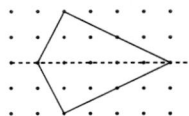

Practice booklet

Sections A and B (p 5)

1 (a) No   (b) Yes   (c) No
   (d) Yes   (e) No   (f) No
   (g) Yes   (h) No   (i) Yes

Section C (p 6)

1 (a) Yes   (b) Yes   (c) Yes
   (d) No   (e) No   (f) No
   (g) Yes   (h) No

Section D (p 7)

1 1881, 1331, 1301, 1811 and 1380 are symmetrical.

2 The pupil's four symmetrical years

3 (a) 1881 has two lines of symmetry.
   (b) The pupil's two years with two lines of symmetry: for example, 1001, 8008, 8118, 1111, 8888

4 (a) One line   (b) No lines
   (c) Two lines   (d) One line

5 (a) 130 (b) 380 (c) 99 (d) 121
The answers to (a) and (b) are symmetrical.

6 The pupil's symmetrical sums

⑥ Multiplication tables

Some guidance on mental methods can be found in the introduction to this guide.

p 39	**A** Up to 5 × 5	Multiplication tables up to 5 × 5 Multiplication by 10
p 41	**B** Up to 10 × 10	Multiplication tables up to 10 × 10
p 43	**C** Tables puzzles	Solving puzzles using multiplication facts

Essential	**Optional**
Two dice, each numbered 0 to 5 Sheet 28	Sheet 36
Practice booklet pages 8 to 10	

🅐 **Up to 5 × 5** (p 39)

Two dice, each numbered 0 to 5, for the caller; sheet 28

To make the learning of tables facts more manageable, this section uses tables up to 5 × 5 and multiplying by 10 only.

You may wish to do questions A1 to A3 before playing 'Tables bingo'.

Tables bingo (p 39)

◊ You can start by giving pupils a free choice of which seven numbers to put on their 'card'. Alternatively, you could ask them to put, say, 12, 15, 25 and four more of their choice.

For weaker pupils, it may be best to give a list of, say, 20 numbers from which they choose. Initially, it could be very demoralising for a pupil to choose 'impossible' results such as 7, 11, 30 etc. Later on they could be given a free choice.

'The pupils all engaged well with this activity. I used it at the end of some lessons to bring the class together.'

◊ The game can easily be varied to include other tables facts. For example, you could number one dice 2, 2, 5, 5, 10, 10 and the other 3, 4, 6, 7, 8, 9.

When pupils are working with sheet 28, encourage them to use multiplication rather than repeated addition.

B **Up to 10 × 10** (p 41)

◊ You could begin the discussion by asking pupils which tables facts they think they know 'by heart' and which they have to think about. Discuss how they work these out. A variety of methods of calculating 6 × 8 are shown and these could support your discussion. Most rely on knowledge of easier tables facts.

Discuss quick methods for other multiplications.

Some possibilities are

• to multiply by 5, multiply by 10 and then halve the result
• to multiply by 4, double and then double again
• to multiply by 6, multiply by 3 and then double the result
• to multiply by 9, multiply by 10 and then subtract

Pupils may be interested in looking at patterns in the 9 times table and this could help their recall.

$9 \times 1 = 9$

$9 \times 2 = 18$

$9 \times 3 = 27$

$9 \times 4 = 36$

$9 \times 5 = 45 ...$

Pupils may be able to use the fact that the first digit of $9 \times n$ is always one less than n and the digits of $9n$ add up to 9 (for $n = 1$ to 10) to help them learn these facts.

◊ Each pupil could be issued with a standard 'tables square' (or write out their own). They could put stickers on it to cover the tables facts they know already and add to it as they learn more. The tables square can be used to show that the number of facts to memorise is less than might be thought at first: if you can put a sticker over, say, 3 × 7 then 7 × 3 can be covered as well.

◊ 'Tables bingo' (see section A) can be extended to cover the majority of tables facts. (Each dice could be numbered 4 to 9.) You could write a list of 20 numbers on the board, from which each pupil picks seven to be their 'card'. (Include some non-starters, like 47, and ask later why they are avoided.)

◊ Instead of the standard 'tables test', you could use a pack of ordinary playing cards (without picture cards) to generate the numbers to be multiplied. The class will feel smug when they get 1 × 2, etc.!

ℂ Tables puzzles (p 43)

In addition to multiplication tables practice, these puzzles provide an opportunity for logical thinking.

Optional: sheet 36

◊ You may wish to use the words 'factor' and 'multiple' in discussing strategies to solve these puzzles.

𝔸 Up to 5 × 5 (p 39)

A1 (a) 8   (b) 25   (c) 38   (d) 43

A2 (a) 20   (b) 15   (c) 30   (d) 42

A3 (a) 15   (b) 16   (c) 15   (d) 60

A4 (a) 18   (b) 37   (c) 44   (d) 27

    (e) 39   (f) 57   (g) 47   (h) 61

A5 Some of these have quite a few possible solutions: no more than two examples are given in each case.

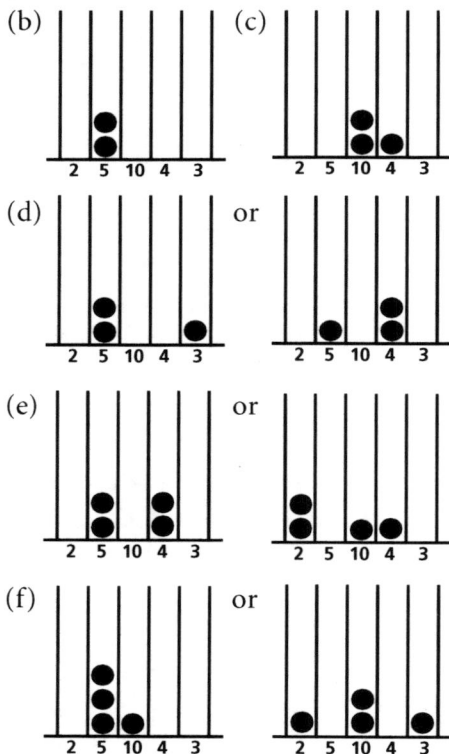

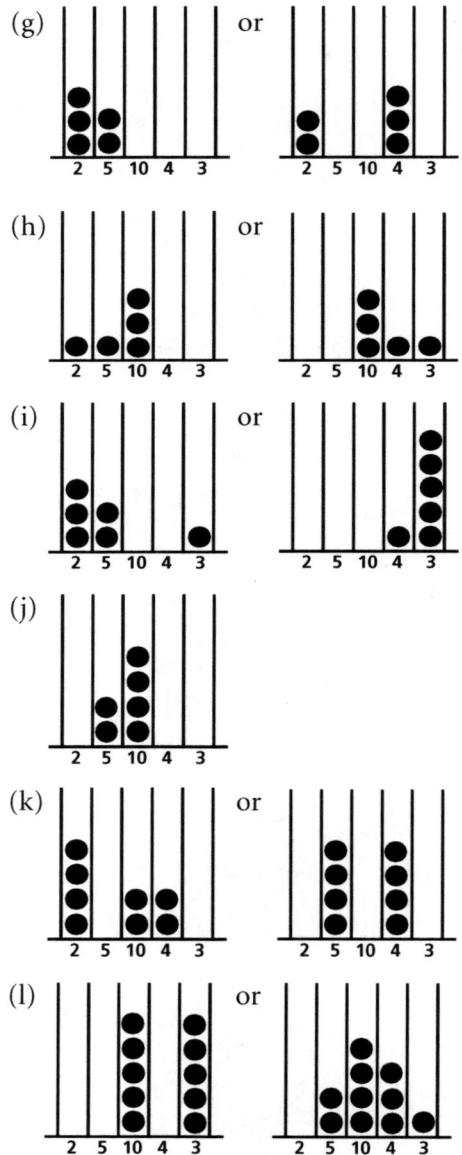

(b)

2 5 10 4 3

(c)

2 5 10 4 3

(d) or

2 5 10 4 3   2 5 10 4 3

(e) or

2 5 10 4 3   2 5 10 4 3

(f) or

2 5 10 4 3   2 5 10 4 3

(g) or

2 5 10 4 3   2 5 10 4 3

(h) or

2 5 10 4 3   2 5 10 4 3

(i) or

2 5 10 4 3   2 5 10 4 3

(j)

2 5 10 4 3

(k) or

2 5 10 4 3   2 5 10 4 3

(l) or

2 5 10 4 3   2 5 10 4 3

B Up to 10 × 10 (p 41)

B1 (a) 35 (b) 24 (c) 28
(d) 27 (e) 80 (f) 40

B2 (a) 42 (b) 64 (c) 56
(d) 54 (e) 49 (f) 36

B3

×	2	0	4
5	10	0	20
3	6	0	12
1	2	0	4

B4 (a)

×	5	2	4
3	15	6	12
1	5	2	4
0	0	0	0

(b)

×	6	3	4
5	30	15	20
3	18	9	12
2	12	6	8

(c)

×	5	6	2
7	35	42	14
10	50	60	20
1	5	6	2

(d)

×	2	4	7
8	16	32	56
4	8	16	28
5	10	20	35

(e)

×	9	3	5
8	72	24	40
5	45	15	25
6	54	18	30

(f)

×	9	7	6
9	81	63	54
8	72	56	48
7	63	49	42

(g)

×	3	5	8
4	12	20	32
1	3	5	8
10	30	50	80

(h)

×	8	5	4
9	72	45	36
6	48	30	24
7	56	35	28

(i)

×	7	6	9
6	42	36	54
3	21	18	27
4	28	24	36

C Tables puzzles (p 43)

C1

×	7	3	2	5
6	42	18	12	30
4	28	12	8	20
9	63	27	18	45
8	56	24	16	40

C2 (a)

×	3	6	5	9
8	24	48	40	72
2	6	12	10	18
4	12	24	20	36
7	21	42	35	63

(b)

×	7	8	5	9
4	28	32	20	36
3	21	24	15	27
6	42	48	30	54
2	14	16	10	18

(c)

×	9	6	7	2
8	72	48	56	16
3	27	18	21	6
4	36	24	28	8
5	45	30	35	10

(d)

×	6	3	2	4
7	42	21	14	28
9	54	27	18	36
5	30	15	10	20
8	48	24	16	32

(e)

×	3	8	5	7
2	6	16	10	14
4	12	32	20	28
9	27	72	45	63
6	18	48	30	42

(f)

×	4	9	6	5
8	32	72	48	40
2	8	18	12	10
7	28	63	42	35
3	12	27	18	15

What progress have you made? (p 43)

1 (a) 6 (b) 12 (c) 10
(d) 15 (e) 20 (f) 9

2 (a)

×	3	5	9
7	21	35	63
10	30	50	90
2	6	10	18

(b)

×	7	8	4
9	63	72	36
8	56	64	32
6	42	48	24

Practice booklet

Section A (p 8)

1 (a) $3 \times 2 = \mathbf{6}$ (b) $2 \times \mathbf{5} = 10$

   (c) $1 \times 3 = \mathbf{3}$ (d) $2 \times 2 = \mathbf{4}$

   (e) $4 \times \mathbf{4} = 16$ (f) $5 \times 4 = 20$

   (g) $\mathbf{0} \times 3 = 0$ (h) $10 \times \mathbf{3} = 10$

2 (a) 15   (b) 20   (c) 12   (d) 14

   (e) 49   (f) 37   (g) 22

3 (a) 8   (b) 15   (c) 10   (d) 12

   (e) 25   (f) 9   (g) 16   (h) 0

4 (a) 30   (b) 48   (c) 60

Sections B and C (p 9)

1 (a)

Table B

×	2	4
3	**6**	**12**
5	**10**	**20**

Table C

×	2	5
3	**6**	**15**
4	**8**	**20**

   (b) Table B 48, table C 49

   (c) Table C

2 (a)

Table A

×	3	5
6	**18**	**30**
8	**24**	**40**

Table B

×	3	6
8	**24**	**48**
5	**15**	**30**

Table C

×	3	8
6	**18**	**48**
5	**15**	**40**

   (b) Table A 112, table B 117, table C 121

   (c) Table C

3 (a) The pupil's tables

   (b) The biggest total is 169.

     An example of a table that gives this total is:

×	4	9
7	28	63
6	24	54

Tables code (p 10)

	(a)	(b)	
1	0	7	W
2	4	9	E
3	2	6	L
4	2	6	L
5	1	9	T
6	4	5	A
7	4	6	B
8	2	6	L
9	4	9	E
10	1	8	S
11	1	9	T
12	3	7	H
13	4	9	E
14	2	5	K
15	2	8	N
16	2	9	O
17	0	7	W
18	0	9	Y
19	2	9	O
20	0	5	U

YOU KNOW THE TABLES WELL

7 Angle

Essential

Sheet 73 (for you: one on white card, one on coloured card)
Sheet 74 (for each pupil: one circle on white card, one on coloured card)
Sheet 75 or 76 or 77 (see below, section D)
Tracing paper
Angle measurers

Practice booklet pages 11 to 13 (angle measurers needed)

𝔸 **Making angles** (p 44)

Use sheet 73 to make a large angle-maker for yourself.
Each pupil needs a small angle-maker, made from:
 • one circle from sheet 74 on coloured card
 • one circle from sheet 74 on white card

If pupils are to make their own, it is better if they follow a demonstration from you. (You can keep them for other groups.)

◊ You can use the scissors pictures to find out what pupils know already about angle. Do they, for example, think that the size of the scissors affects the size of the angle?

Using the angle-maker

◊ Turn the coloured circle gradually to make each quarter turn like this.

Ask pupils to describe each of the coloured angles in as many ways as they can. Record the words they use and link them together. For example, 90° = right angle = quarter turn.

Set the angle-maker to zero again, and start with the line neither vertical nor horizontal. Ask pupils to tell you to stop turning when you have made

- a right angle
- a half turn
- three quarters of a turn

Some pupils find it difficult to recognise these angles in different orientations.

◊ Introduce the terms acute, obtuse and reflex.

You could define them in terms of quarter turn, half turn, etc. (unless pupils are already familiar with degrees). Make some more angles in various orientations for pupils to describe as acute, obtuse or reflex.

Comparing angles

◊ Some pupils see this as a 'large' angle:

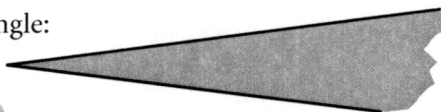

and this as a 'small' angle:

The purpose of the following work is to emphasise that the angle is the amount of turn and has nothing to do with the lengths of the arms or the area between them.

Show an angle on your angle-maker. Ask pupils to make it on their smaller ones. Invite some of them to compare their angle to yours by placing it over yours. Emphasise that the length of the arms doesn't matter.

> 'I did do this and it worked very well. I didn't spend too much time on it as almost all of the class knew quite a bit about angles already.'

Make an angle on a large angle-maker and another on a small angle-maker. Ask which is the bigger angle. Include examples where

- the larger angle is on the smaller angle-maker
- both angles are the same
- the larger angle is on the larger angle-maker
- the two angles are held in different orientations

Describing angles

◊ Pupils work in pairs, sitting back to back. One makes an angle and describes it as closely as they can. The other makes the angle from the description. Then they compare the angles and see how close they were.

B Comparing angles (p 45)

Tracing paper

This work can help you identify pupils who are still unsure about what makes one angle bigger than another.

C Right angles, acute, obtuse and reflex angles (p 46)

D Measuring angles (p 48)

Angle measurers (360°), sheets 75, 76, 77 (graded in difficulty; see D1 below)

◊ Use an OHP to demonstrate how to use the angle measurer. Show that you can measure starting from either arm. Discuss which scale to use. Link degrees to fractions of a turn.

◊ Draw some angles. Ask pupils to estimate then measure each of them.

D1 The three sheets are graded as follows:

- sheet 75   angles are all multiples of 5° (point this out to pupils)
- sheet 76   angles are of any size
- sheet 77   angles are marked in the corners of polygons

You may wish pupils to use more than one of these sheets.

Tilting bus

This is intended to stimulate discussion. The pointer on the body shows how far the body has tilted. The other pointer shows how far the chassis has tilted. The difference is due to the 'give' in the suspension system.

A double decker bus must remain stable when the chassis is tilted to 28° (in practice it can go much further than this). The test has to be done with weights added to simulate a full load of passengers upstairs but no extra weight downstairs.

E Drawing angles (p 50)

Angle measurers

T

◊ Some pupils are reluctant to extend lines beyond the point they have marked against the angle measurer.

F Angles on a line (p 51)

Angle measurers

T

◊ Measurement errors may need to be dealt with in the initial activity.

B Comparing angles (p 45)

B1 Angles *a* and *c* are bigger than *X*.
Angles *b* and *d* are smaller than *X*.

B2 (a) Bigger   (b) Bigger   (c) Smaller
(d) Bigger   (e) Bigger

B3 *c*, *a*, *b*, *d*

C Right angles, acute, obtuse and reflex angles (p 46)

C1 *a* and *c* are right angles.

C2 (a) Acute   (b) Right angle
(c) Acute   (d) Obtuse
(e) Obtuse

C3 *a* is a right angle,   *b* is acute,
c is acute,   *d* is obtuse,
e is acute,   *f* is acute.

C4 *a* is acute,   *b* is obtuse,
c is acute,   *d* is reflex,
e is obtuse,   *f* is a right angle.

C5 *a* is acute,   *b* is a right angle,
c is reflex,   *d* is a right angle,
e is obtuse,   *f* is reflex,
g is acute.

D Measuring angles (p 48)

D1 Pupils' answers should be within 1° of the answers given.

Sheet 75

$a = 30°$, acute $b = 40°$, acute
$c = 55°$, acute $d = 135°$, obtuse
$e = 15°$, acute $f = 265°$, reflex
$g = 90°$, right angle $h = 235°$, reflex

Sheet 76

$a = 15°$, acute	$b = 90°$, right angle
$c = 50°$, acute	$d = 93°$, obtuse
$e = 123°$, obtuse	$f = 212°$, reflex
$g = 270°$, reflex	$h = 117°$, reflex

Sheet 77

$a = 70°$, acute	$b = 60°$, acute
$c = 107°$, obtuse	$d = 73°$, acute
$e = 107°$, obtuse	$f = 47°$, acute
$g = 112°$, obtuse	$h = 56°$, acute
$i = 60°$, acute	$j = 60°$, acute

D2 a is equal to e; b is equal to f.

D3 (a) $x = 33°$, $y = 28°$, $z = 119°$
  (b) They add up to 180°.
  (c) You should get the same total for any triangle.

D4 (a) $a = 100°$, $b = 139°$, $c = 48°$, $d = 73°$.
  (b) They add up to 360°.
  (c) You should get the same total for any four-sided shape.

E Drawing angles (p 50)

E1 90°

E2 50°

E3 40°

E4 (a) 80°   (b) 360°

F Angles on a line (p 51)

F1 $a = 140°$ $b = 35°$ $c = 130°$ $d = 20°$

F2 $a = 40°$ $b = 60°$ $c = 115°$

F3 $a = 39°$ $b = 43°$ $c = 97°$

What progress have you made? (p 52)

1 a is an acute angle, b is a right angle, c is an acute angle, d is an obtuse angle.

2 a, c, b, d

3 $a = 50°$ $b = 90°$ $c = 84°$ $d = 136°$

4 80°

5 $a = 50°$ $b = 70°$

Practice booklet

Sections A, B and C (p 11)

1 b, a, d, c

2 p is obtuse, q is a right angle,
  r is acute, s is acute.

3
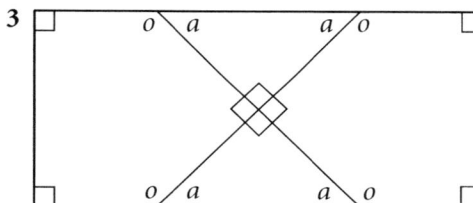

Section D (p 12)

Essential: angle measurer

1 $a = 32°$ $b = 61°$ $c = 128°$
  $d = 90°$ $e = 14°$

Sections E and F (p 13)

1 (a) 40°   (b) 360°

2 $a = 130°$ $b = 105°$ $c = 60°$
  $d = 25°$ $e = 80°$ $f = 140°$
  $g = 65°$ $h = 34°$

⑧ Oral questions: calendar (p 53)

This page is for teacher-led oral work on interpreting a calendar.

Using oral activities

◊ Aim for sessions of oral questions that are regular and fairly short, with all pupils feeling a sense of achievement at the end. It is intended that you can use the pages more than once. Keep the questions simple when doing this work for the first time. Do not give too many questions: it is better to stop with all pupils feeling some sense of achievement after, say, five minutes rather than persist for half an hour!

◊ An initial discussion of pupils' own calculating methods may be beneficial, especially ways of calculating mentally with money.

◊ Oral work of this kind is usually most successful when pupils work in silence. Explain that you will read each question twice and then they will be given time to answer it. Some schools allowed 15 seconds between questions, others allowed more. Emphasise that pupils need to listen very carefully and if they cannot do a question they should forget it and try the next one.

You may like to record the questions on a tape with a specified time limit between questions and play that in class. This can encourage pupils to concentrate on listening to the questions.

◊ Some possible questions are given for each oral activity. You can make up your own. You could also ask each pupil to write one question, and then give all the questions orally in class.

◊ These sample questions are roughly in order of increasing difficulty.

1 What day of the week is 3 March in the year 2011?   Thursday

2 Is 2011 a leap year?   No

3 How many days are there in September?   30

4 What day of the week is Christmas day in 2011?   Sunday

5 How many Saturdays are there in October in 2011?   5

6 Which months have five Thursdays in them?

March, June, September, December

7 Whitsun bank holiday is the last Monday in May.
   What date is that?   30 May

8 The first day of a school trip is 22 June and the last day is 30 June. How many days is that?   9

9 How many shopping days (not Sundays) are there from 21 November until Christmas?   30

10 Karen sends a postcard on 9 August. It gets to her grandmother four days later. What date does it get there?   13 August

11 Ted gets a bill dated 5 October and has 21 days to pay. What is the latest date he can pay?   26 October

12 On 5 August, Cyril gets a parcel that was posted in Australia on 13 July. How many days did it take to arrive?   23

13 Peter gets some books out of the library on 27 May. They are due back three weeks later. What date are they due back?   17 June

14 The last day of Nicola's school term is 22 July. The first day back in the autumn is 5 September. How many weeks summer holiday does she get?   6

15 Ted hires some scaffolding on Monday 24 October. It is due back to the hire company on the Monday six weeks later. What date is it due back?   5 December

16 Councillor James holds a residents' advice session on the third Wednesday in every month. What dates does he hold it in July and August?   20 July, 17 August

17 Mrs Brown's home help comes on alternate Wednesdays, starting on the fifth of January. What dates does she come in March?   2, 16, 30 March

You can make work of this kind more immediate by working from an OHP photocopy of a current calendar and referring to actual dates of school events, sports fixtures and so on.

⑨ Growing patterns

Pupils investigate sequences arising from a variety of contexts.

The emphasis is on finding a rule to continue a sequence and explaining why the rule is valid. There is a little work on finding a rule that effectively gives the nth term (though not expressed in that way), but this aspect is covered more fully in unit 24 'Work to rule'.

Pupils should realise that just spotting a pattern in the first few numbers in a sequence (for example, 'add 3 to the previous number') is not enough to prove that the sequence will continue in the same way. You have to go back to the context (rose bushes, ponds, earrings, ...) to give a convincing explanation.

p 54	**A** Coming up roses	Investigation leading to a linear sequence Describing how the sequence continues Explaining why the sequence continues like this
p 54	**B** Pond life	Investigation leading to a linear sequence
p 56	**C** Changing shape	Consolidating work on linear sequences
p 57	**D** Earrings	Investigation leading to a simple exponential sequence (powers of 2)

Optional
Coloured counters
Coloured tiles
Multilink cubes
Square dotty paper
Triangular dotty paper
Sheet 70

Practice booklet pages 14 and 15

68 • 9 Growing patterns

A Coming up roses (p 54)

This investigation gives rise to a linear sequence.
Pupils describe and explain how the sequence continues.

> Optional: tiles/counters to represent red and white rose bushes

◊ A gardener is designing a display with red and white rose bushes planted as single rows of red roses, with a row of white roses on each side and a white rose at each end. Some examples are:

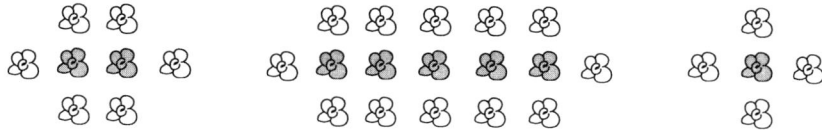

represents a white rose bush.

represents a red rose bush.

◊ To start with, pupils could think about the designs on the page and then try to draw similar arrangements that use 2, 6, 4 red roses, for example.

Encourage pupils to simplify the diagrams, for example by using coloured circles or the letters R and W.

Pupils count the number of red and white rose bushes needed in each arrangement so far and collect their results together in an ordered table. Discuss the advantages of tabulating in this way.

Alternatively, pupils could consider what the smallest design would look like and draw it. A sequence of designs can then be produced in order: 1 red rose bush, 2 red rose bushes etc. These results can then be tabulated.

◊ Ask pupils to complete the table up to, say, 8 red rose bushes. Discuss any methods that pupils use to complete the table. These may include

 • making or drawing the designs
 • finding and using the rule that the number of white rose bushes increases by 2 for every extra red rose bush
 • finding and using the rule that the number of white rose bushes is (2 × *the number of red rose bushes*) + 2

It is important that each method is considered equally valid.

◊ In discussion, bring out the fact that the number of white rose bushes increases by 2 for every extra red rose bush. Ask the pupils to use the diagrams to explain why this is so.

You could ask the pupils to consider a range of 'explanations'.
For example, the number of white rose bushes increases by 2 for every extra red rose bush because

- the numbers in the table go up in 2s
- the row of red bushes has 1 white bush at each end (2 in total)
- there are 2 colours of roses
- each red rose bush has 1 white rose bush either side of it (2 in total)

These statements could be put on cards and groups of pupils could consider which is an acceptable explanation.

Many pupils would think that just describing how the sequence continues (as in the first statement above) is a perfectly acceptable 'explanation'. Emphasise they must refer back to the arrangement of bushes to explain why the sequence will continue in the same way.

◊ Now ask pupils to think about a larger number of red bushes, for example 20 red bushes, and to say how many white bushes would be needed for them. Again, pupils are likely to use a variety of methods as before.

If pupils use the rule that the number of white rose bushes is (2 × the number of red rose bushes) + 2, ask them to explain why they know their rule works by referring to the arrangement of rose bushes. Emphasise that just because their rule works for a few results it does not follow it will work for all results.

Pose a question like '314 white rose bushes are needed for 156 red rose bushes. So how many will be needed for 157 rose bushes?' and ask pupils to discuss their methods.

ⓑ **Pond life** (p 54)

Pupils investigate another situation that gives rise to a linear sequence. They describe and explain how the sequence continues.

Optional: square dotty paper, coloured tiles

◊ Make sure pupils are clear that the ponds are square ponds.

B5 A variety of methods are possible. If pupils are struggling ask them to look at their answer for B4(d) and think how that could help them. Some pupils will continue to draw ponds at this stage.

B7 In part (b), some pupils may say 'because the numbers in the table go up in 4s'. Emphasise that they must refer back to the arrangement of slabs to explain why they can be sure that the sequence continues in the same way.

B8 Ask those who use the 'multiply by 4 and add 4' rule or the 'add 1 and multiply by 4' rule to consider if there is any advantage in using the 'going up in 4s' rule here (the calculation 272 + 4 is simpler than (68 × 4) + 4).

C Changing shape (p 56)

This consolidates work on linear sequences.

Optional: square dotty paper, triangular dotty paper, coloured tiles

C1 In part (c), emphasise that the width of all the ponds for this table is 3 metres. The length of the pond is the other dimension so, for some ponds, the length is shorter than the width.

As an extension pupils could consider triangular ponds surrounded by triangular slabs.

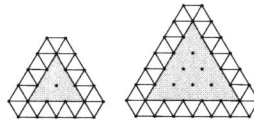

D Earrings (p 57)

Pupils investigate a situation leading to a simple exponential sequence (powers of 2). They find and explain how the sequence continues.

Optional: red and yellow multilink cubes, sheet 70 (for recording results)

◊ Multilink has been found to be a very useful way to 'build' the earrings. It allows easy identification of duplicates and collection of results.

If pupils use multilink, make sure they realise that these two designs are different.

◊ Ask pupils to find as many different three-bead earrings as they can.

Collect the results for the whole class and discuss how they can be sure that they have found all possible designs for three beads.

The 8 different designs are:

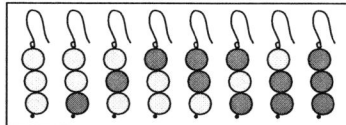

◊ It is important to stress that 'predict and check' is a good strategy to increase confidence that any patterns or rules found are correct. However, predict and check does not explain *why* any patterns or rules are valid and hence is not a proof that the sequence of numbers will continue in the way they expect.

D3 In part (c), pupils can record their results on sheet 70.

D4 Pupils cannot claim to be sure about the number of five-bead earrings until they have found them all (and shown no more exist) or until they have explained why the numbers in the sequence double each time.

B Pond life (p 54)

B1 (a) 8 slabs   (b) 12 slabs   (c) 16 slabs

B2 (a)

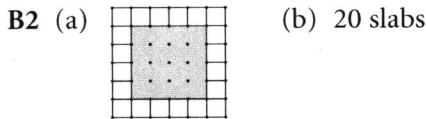

(b) 20 slabs

B3 (a)

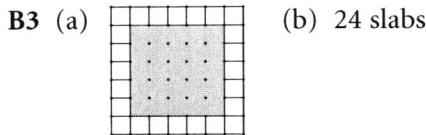

(b) 24 slabs

B4 (a) The numbers of slabs in the table are 8, 12, 16, 20, 24, 28.

(b) 32 slabs   (c)

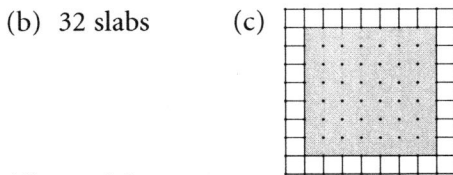

(d) 44 slabs
Pupils' methods are likely to involve
- counting on in 4s or
- multiplying by 4 and adding 4 or
- adding 1 and multiplying by 4

B5 (a) 11 metres
Pupils' methods could involve

- extending the pattern in the table or
- working from the fact that a 10 by 10 pond needs 44 slabs or
- subtracting 4 and dividing by 4 or
- dividing by 4 and subtracting 1

(b) A 14 metre pond

B6 (a) 4 extra   (b) 68 slabs

B7 (a) The number of slabs needed goes up by 4 each time.

(b) The pupil's explanation: for example, an increase of 1 metre in width means an extra slab for each edge. Since the pond has 4 edges, 4 extra slabs are needed.

B8 276 slabs

C Changing shape (p 56)

C1 (a) 20 slabs   (b) 14 slabs

C2 (a)

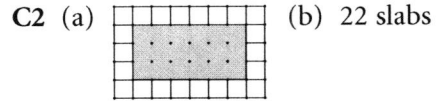

(b) 22 slabs

C3 (a) 24 slabs   (b)

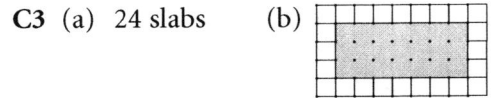

C4 (a) The numbers of slabs in the table are 12, 14, 16, 18, 20, 22, 24.

(b) 26 slabs
The pupil's method: for example, 24 + 2 = 26

C5 (a) The number of slabs needed goes up by 2 each time.

(b) The pupil's explanation

D Earrings (p 57)

D1 4 different earrings

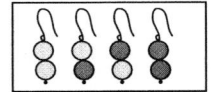

D2 The numbers of different earrings in the table are 2, 4, 8.

D3 (a) 16 different earrings

(b) The pupil's method: for example, 8 × 2 = 16 or 8 + 8 = 16

(c) One way to organise the results is to add a yellow bead to each of the three-bead earrings and then add a red bead.

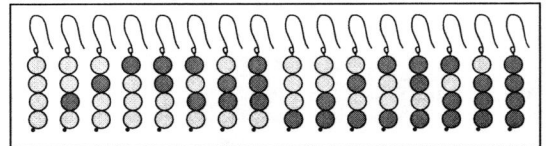

Another way is to look at earrings with 0 red beads, 1 red bead, 2 red beads, …

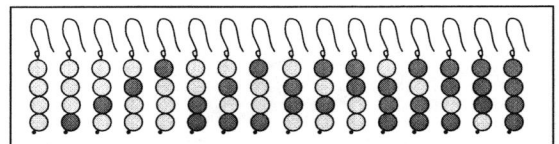

D4 32 different earrings

D5 The number of earrings doubles each time.

What progress have you made? (p 58)

1 (a) 6 bushes   (b) 8 bushes

2 🌸 🌸 🌸 🌸 🌸
    🌸 🌸 🌸 🌸 🌸

3 The numbers of yellow rose bushes in the table are 5, 6, 7, 8, 9.

4 (a) 12 yellow bushes

    (b) The pupil's method: for example, count in 1s from the result for 5 red bushes.

5 (a) The number of yellow bushes goes up by 1 each time.

    (b) The pupil's explanation: for example, an increase of 1 red bush means an extra yellow bush in the bottom row.

6 (a) 14 yellow bushes

    (b) The pupil's method: for example, $10 + 4 = 14$

Practice booklet

Sections A, B and C (p 14)

1 The pupil's check

2 7 corners

3 The pupil's chain of 4 squares; 13 corners

4

Number of squares	1	2	3	4
Number of corners	**4**	**7**	10	**13**

5 (a) 16 corners

    (b) The pupil's drawing of a 5-square chain

6 31 corners

7 (a) The number of corners goes up by 3 for each added square.

    (b) The corner of a new square that joins the chain has already been counted, so 3 corners are added each time.

8 61 corners

Section D (p 15)

1 The pupil's drawing for year 4; 8 flowers

2

Year	1	2	3	4
Number of flowers	1	2	**4**	**8**

3 16 flowers

4 The pupil's drawing for year 5, with 16 flowers

5 The number of flowers doubles each year.

6 Each year every flower from the year before is replaced by two flowers.

Two-piece tangrams (p 59)

These activities give an opportunity to use mathematical language for shapes, properties etc., and to encourage good presentation. Because the possibilities are limited in each case, it is fairly easy to find them all.

Essential	Optional
Squared paper, scissors, glue	Card
	Pre-drawn rectangles (see below)

1 Rectangle (p 59)

◊ The dimensions 7 cm and 4 cm are chosen because an equilateral triangle (within the bounds of measurement error!) can be made with the two pieces.

◊ Instead of making several copies, pupils could cut one pair of triangles from card and draw round the shapes they make.

The pieces may be turned over to make the shapes.

◊ These are the different shapes that can be made:

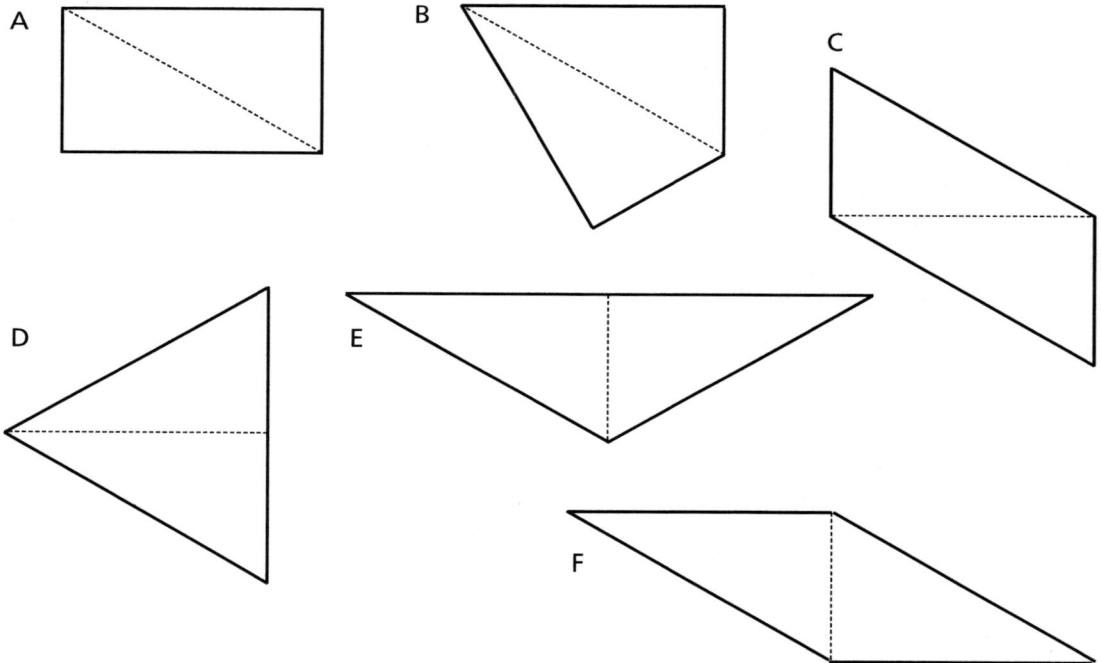

Follow-up

You could write the names of the shapes on the board (kite, isosceles triangle etc.) and ask pupils to label their shapes.

Pupils could suggest ways of categorising. The results can be tabulated, for example:

	Sides	Name	Right angles	Lines of symmetry
A	4	rectangle	4	2
B	4	kite	2	1

2 Square (p 59)

These shapes can be made:

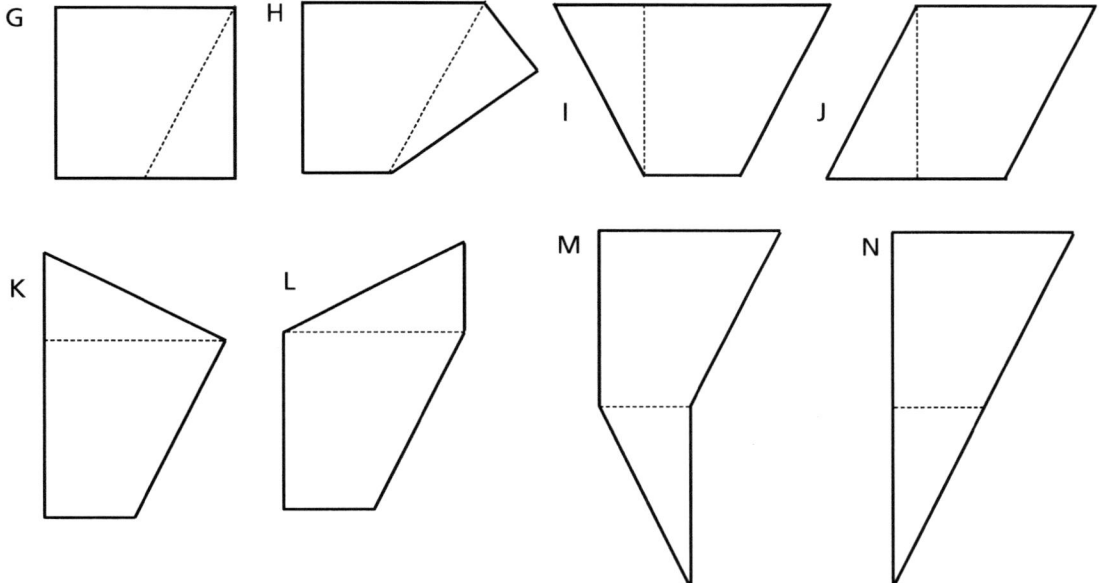

Extensions

◊ Pupils could start with their own shape and make a two-piece tangram.

◊ Add an extra cut to the rectangle or square to make a three-piece tangram. Suggestions (if needed):

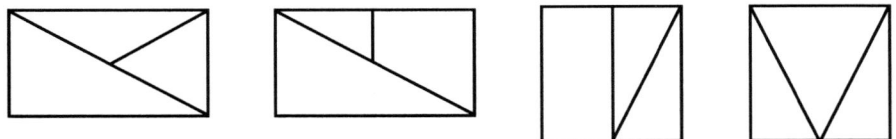

It is better if the cut gives equal lengths for joining.

⑪ Fractions

Essential

Sheets 36 and 37 (squared paper can be used instead of sheet 36), square dotty paper

Practice booklet pages 16 to 18

Ⓐ Fractions everywhere (p 60)

◊ Before starting, you could ask pupils what they already know about fractions. They can discuss this in pairs before you collect ideas together.

The pictures on pages 60 and 61 are intended to stimulate discussion about the meaning of simple fractions used in everyday life. Ask pupils to be as precise as possible when they explain what each fraction means.

Ⓑ Halves, quarters and what else? (p 62)

◊ Pupils could work in pairs first, with a class discussion to follow. Focus on pupils' explanations of their answers. Some may be able to see when half or a quarter is coloured but may not be able to describe other fractions. This does not matter at this stage: it is dealt with in later sections.

C introduces the idea that a fraction may be made up of separated parts. Pupils who say P is $\frac{1}{2}$ coloured and K and N are $\frac{1}{4}$ coloured are counting parts regardless of whether they are equal.

The shapes that are half coloured are A, B, C, E and F.

Those that are a quarter coloured are D, H and I (the more able may see that P is also).

The more obvious other fractions are G $\frac{1}{3}$, K $\frac{1}{6}$, L $\frac{3}{4}$, M $\frac{1}{16}$ and J $\frac{1}{6}$.

N $\frac{1}{8}$ and O $\frac{1}{8}$ are much less obvious and it may only be the more able who see them.

Ⓒ Shade half, shade a quarter (p 63)

Sheet 36 or squared paper

◊ Pupils need only do a few of these to get the point. More confident pupils can include diagonal lines.

Ⓓ Thirds and what else? (p 63)

Square dotty paper, sheet 37

E **Eighths** (p 65)

◊ You may wish to introduce this using folded paper.

C **Shade half, shade a quarter** (p 63)

C1 and **C2** The pupil's shaded squares

D **Thirds and what else?** (p 63)

D1 The pupil's shaded rectangles

D2 A and C

D3 The pupil's shaded squares

D4 A and C

D5 (a) $\frac{1}{3}$ (b) $\frac{2}{3}$ (c) $\frac{3}{4}$ (d) $\frac{1}{3}$ or $\frac{3}{9}$
(e) $\frac{1}{3}$ or $\frac{2}{6}$ (f) $\frac{1}{4}$ (g) $\frac{1}{9}$ (h) $\frac{3}{4}$

D6 A The pupil's shading in of the shapes
B (a) $\frac{1}{4}$ (b) $\frac{2}{3}$
   (c) $\frac{1}{2}$ (or $\frac{2}{4}$) (d) $\frac{1}{3}$ (or $\frac{2}{6}$)
   (e) $\frac{3}{4}$ (or $\frac{6}{8}$) (f) $\frac{1}{3}$ (or $\frac{3}{9}$)
C (a) The pupil's attempt
       The accurate mark should be
       4 cm from A.
   (b) The pupil's attempt
       The accurate mark should be
       3 cm from C.

E **Eighths** (p 65)

E1 (a) $\frac{4}{8}$ or $\frac{2}{4}$ or $\frac{1}{2}$ (b) $\frac{2}{8}$ or $\frac{1}{4}$
(c) $\frac{6}{8}$ or $\frac{3}{4}$ (d) $\frac{4}{8}$ or $\frac{2}{4}$ or $\frac{1}{2}$
(e) $\frac{4}{8}$ or $\frac{2}{4}$ or $\frac{1}{2}$ (f) $\frac{6}{8}$ or $\frac{3}{4}$

E2 (a), (b) Rectangle with $\frac{3}{4}$ shaded
(c) Rectangle with $\frac{7}{8}$ shaded
(d) $\frac{7}{8}$ is bigger

E3 (a) $\frac{3}{8}$ (b) $\frac{5}{8}$ (c) $\frac{3}{4}$

E4 $\frac{1}{4}$, $\frac{3}{8}$, $\frac{1}{2}$, $\frac{5}{8}$

What progress have you made? (p 66)

1 (a) $\frac{1}{2}$ (b) $\frac{1}{2}$ or $\frac{2}{4}$ (c) $\frac{2}{3}$ (d) $\frac{1}{4}$
(e) $\frac{3}{4}$ (f) $\frac{3}{4}$

2 (a) $\frac{1}{4}$ or $\frac{2}{8}$ (b) Two of $\frac{4}{8}$ or $\frac{2}{4}$ or $\frac{1}{2}$

3 (a) Rectangle with $\frac{3}{8}$ shaded
(b) Rectangle with $\frac{3}{4}$ shaded
(c) $\frac{3}{4}$ is bigger

4 $\frac{1}{4}$, $\frac{1}{2}$, $\frac{3}{4}$, $\frac{7}{8}$

Practice booklet

Sections B, C and D (p 16)

1 (a) $\frac{1}{2}$ (b) $\frac{1}{4}$ (c) $\frac{2}{4}$ or $\frac{1}{2}$ (d) $\frac{3}{4}$

2 The statement is wrong.
The parts are not the same size.

3 (a) Yes (b) Yes (c) No (d) Yes
(e) Yes (f) Yes (g) Yes (h) No

4 (a) Yes (b) Yes (c) No (d) Yes

5 (a) Yes (b) No (c) Yes (d) Yes
(e) Yes (f) Yes

6 (a) Yes (b) Yes (c) No (d) No
(e) No (f) No

Section E (p 18)

1 (a) $\frac{4}{8} = \frac{2}{4} = \frac{1}{2}$ (b) $\frac{6}{8} = \frac{3}{4}$
(c) $\frac{2}{8} = \frac{1}{4}$ (d) $\frac{4}{8} = \frac{2}{4} = \frac{1}{2}$

2 (a) The pupil's shading of $\frac{1}{2}$ of the square
(b) The pupil's shading of $\frac{3}{4}$ of the square
(c) The pupil's shading of $\frac{5}{8}$ of the square
(d) $\frac{1}{2}$, $\frac{5}{8}$, $\frac{3}{4}$

3 (a) $\frac{1}{8}$ (b) $\frac{3}{8}$ (c) $\frac{3}{4}$

4 Richard eats more; $\frac{3}{8}$ is larger than $\frac{1}{4}$.

5 $\frac{3}{8}$, $\frac{1}{2}$, $\frac{5}{8}$, $\frac{3}{4}$, $\frac{7}{8}$

Review 1 (p 67)

An angle measurer is needed for question 11.

1 (a)   (b)

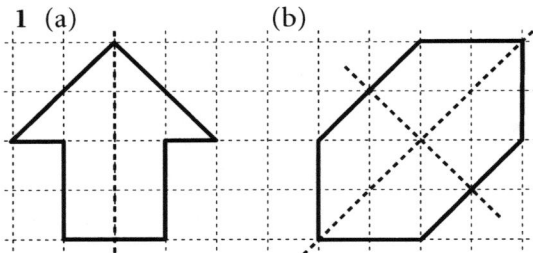

2 (a) Line 2   (b) Lines 1 and 3

3 (a) 10 (b) 20 (c) 72 (d) 97

4 (a)

×	3	7	4	5
2	6	14	8	10
3	9	21	12	15
6	18	42	24	30
10	30	70	40	50

(b)

×	6	3	4	8
5	30	15	20	40
6	36	18	24	48
4	24	12	16	32
7	42	21	28	56

(c)

×	10	8	6	5
7	70	56	42	35
4	40	32	24	20
3	30	24	18	15
9	90	72	54	45

5 (a) 16   (b) 8

6 The pupil's drawing for 2 purple tulips and 10 yellow tulips.

7

No. of purple tulips	1	2	3	4	5	6
No. of yellow tulips	8	10	12	14	16	18

8 (a) 26   (b) 12

9 (a) The number of yellows goes up in 2s.

    (b) The pupil's explanation

10 (a) a and d are right angles, b is obtuse and c is acute.

     (b) $a = 90°$, $b = 145°$, $c = 35°$, $d = 90°$

11 (a) The pupil's drawing

     (b) 85°   (c) 540°

12 $a = 110°$ $b = 15°$ $c = 155°$

    $d = 25°$ $e = 125°$

13 (a) $\frac{3}{9}$ or $\frac{1}{3}$ (b) $\frac{6}{9}$ or $\frac{2}{3}$

     (c) $\frac{2}{4}$ or $\frac{1}{2}$ (d) $\frac{4}{8}$ or $\frac{2}{4}$ or $\frac{1}{2}$

14 (a) The pupil's square with $\frac{1}{8}$ shaded

     (b) The pupil's square with $\frac{5}{8}$ shaded

     (c) The pupil's square with $\frac{3}{4}$ shaded

     (d) $\frac{3}{4}$ is bigger

15 $\frac{4}{8} = \frac{2}{4} = \frac{1}{2}$

16 $\frac{1}{2}$, $\frac{5}{8}$, $\frac{3}{4}$, $\frac{7}{8}$

Mixed questions 1 (Practice booklet p 19)

1 (a) 217 kilometres

    (b) 23 kilometres

    (c) Going through Bigton; 5 kilometres

2 (a) MUM, BOB, TUT, OXO

    (b) OXO

    (c) The pupil's four words with reflection symmetry

3 (a)

×	5	4	3
2	**10**	8	**6**
7	**35**	**28**	**21**
4	**20**	**16**	**12**

(b)

×	6	**3**	4
5	**30**	**15**	**20**
7	**42**	**21**	**28**
8	**48**	24	**32**

(c)

×	**7**	4	8
3	21	**12**	**24**
5	**35**	**20**	40
9	**63**	**36**	**72**

4 *a* is a right angle, *b* is obtuse,
c is obtuse, *d* is acute,
e is reflex, *f* is acute.

5 $a = 55°$ $b = 110°$
$c = 85°$ $d = 15°$

6 (a) (i) 3 rods (ii) 5 rods
(iii) 7 rods

(b) The pupil's drawing with 5 triangles;
11 rods

(c)

No. of triangles	1	2	3	4	5
No. of rods needed	**3**	**5**	**7**	9	**11**

(d) 13 rods
The pupil's diagram to check

(e) Each time a triangle is added the
number of rods goes up by 2.

(f) 41 rods

7 (a) $\frac{5}{8}$ (b) $\frac{3}{8}$ (c) $\frac{1}{4}$

Essential	**Optional**
Sheets 67, 68 and 69	OHP transparency of sheet 67
Practice booklet pages 21 and 22	

Ⓐ **Recording positions** (p 70)

> Sheet 67 (an OHP is also useful)

◊ The two class games described below can be played at any time.

Coordinate bingo

Each pupil draws a grid labelled 0 to 5 on each axis. They mark seven grid points with little circles. This is their bingo card. You have a grid as well and call out points at random.

Four in a line

Draw a grid on the board labelled 0 to 6 on each axis. Choose two players who take turns to say the coordinates of a point. Label the points with the players' initials. The first to get four of their points in a line is the winner.

For a more demanding, but more interesting, game make it five in a line on a 0 to 9 grid.

Ⓑ **Digging deeper** (p 73)

> Sheet 68 (whole-number coordinates), sheet 69 (includes $\frac{1}{2}$ and 0.5)

◊ Question B1 is for practice in whole-number coordinates.
Pupils who do not need this can start at B2.

Ⓐ Recording positions (p 70)

A1 (a) Boy's boot   (b) Silver brooch
     (c) Wooden comb

A2 (a) $(3, 7)$ (b) $(7, 10)$ (c) $(6, 3)$

A3 A wall

A4 $(0, 0)$

A5 A $(3, 3)$, B $(8, 8)$

A6

Sheet 67

Ⓑ Digging deeper (p 73)

B1

Sheet 68

B2

Sheet 69

B3

Sheet 69

What progress have you made? (p 74)

1 A $(0, 3)$, B $(4, 1)$, C $(6, 0)$

2

Practice booklet

Section A (p 21)

1 (a) SMILE   (b) WITH
   (c) YOUR   (d) FRIENDS

2 (a) $(3, 1)$ $(0, 4)$ $(4, 1)$ $(0, 0)$ $(2, 1)$
       $(3, 4)$ $(0, 4)$ $(4, 0)$
   (b) $(3, 2)$ $(3, 3)$ $(1, 3)$ $(2, 3)$ $(4, 1)$

3
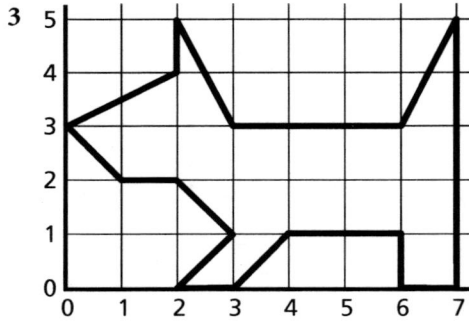

Section B (p 22)

1 (a) $(0, 2)$, $(3, 4)$, $(6, 2)$, $(3, 0)$
   (b) $(3, 2)$

2 (a) $(4, 5)$, $(4, 8)$, $(7, 8)$, $(7, 5)$
   (b) $(5\frac{1}{2}, 6\frac{1}{2})$ or $(5.5, 6.5)$

3 (a) $(7\frac{1}{2}, \frac{1}{2})$, $(7\frac{1}{2}, 4\frac{1}{2})$, $(9\frac{1}{2}, 4\frac{1}{2})$, $(9\frac{1}{2}, \frac{1}{2})$
   (b) $(8\frac{1}{2}, 2\frac{1}{2})$

4 (a) Inside A   (b) Inside C
   (c) Inside B   (d) Not inside any
   (e) Inside A   (f) Inside A

⑬ Area and perimeter

Most of the work involves rectangles. There is introductory work on right-angled triangles.

p 75	**A** Exploring perimeters	
p 77	**B** Square centimetres	Wholes and halves
p 78	**C** The area of a rectangle	Sides in whole centimetres
p 79	**D** Measuring to find areas	
p 81	**E** It pays to advertise	Working out costs that depend on area
p 83	**F** Square metres	Whole numbers only

Essential
Centimetre squared paper

Practice booklet pages 23 to 26

𝔸 Exploring perimeters (p 75)

◊ These investigations are in order of in increasing difficulty. Pupils can go as far with them as they can manage, but all should become familiar with the word *perimeter* and should come to see that a particular number of squares (producing shapes of a fixed area) can give rise to different perimeters. The perimeters are all even numbers of centimetres (investigation 3) because before the squares are put together their total perimeter is an even number of centimetres (actually a multiple of 4) and each time two edges are put together the total perimeter falls by 2 centimetres, remaining even.

The chart on the next page shows the possible perimeters for investigations 2, 4 and 5.

'Students discussed the compactness of the shapes.'

The maximum perimeter for a given number of squares n is $2n + 2$ or $2(n + 1)$ and arises when all the squares are in a straight line or form a shape 'one square wide' with bends in it (the least 'compact' arrangement). Some pupils should be able to explain, in their own words, why such a relationship applies. This linear relationship is shown by the upper right-hand edge of the block of ticks on the chart.

Perimeter / Number of squares chart

Number of squares	4	6	8	10	12	14	16	18	20	22	24	26	28	30	32	34	36
1	✓																
2		✓															
3			✓														
4			✓	✓													
5				✓	✓												
6				✓	✓	✓											
7					✓	✓	✓										
8					✓	✓	✓	✓									
9					✓	✓	✓	✓	✓								
10						✓	✓	✓	✓	✓							
11						✓	✓	✓	✓	✓	✓						
12						✓	✓	✓	✓	✓	✓	✓					
13							✓	✓	✓	✓	✓	✓	✓				
14							✓	✓	✓	✓	✓	✓	✓	✓			
15							✓	✓	✓	✓	✓	✓	✓	✓	✓		
16							✓	✓	✓	✓	✓	✓	✓	✓	✓	✓	
17								✓	✓	✓	✓	✓	✓	✓	✓	✓	✓

Pupils may be able to give their own explanations of why the steps shown with arrows in the chart occur.

B Square centimetres (p 77)

C The area of a rectangle (p 78)

D Measuring to find areas (p 79)

◊ Some pupils may need support with questions D3 and D4, where the process is inverted.

E It pays to advertise (p 81)

◊ This is straightforward practice.

F Square metres (p 83)

◊ Pupils often calculate answers in square metres with no idea how big a square metre is. At this stage, it is worthwhile making a metre square from newspaper and referring to it in working out or estimating the area of, say, the classroom wall.

It is also worth doing some work on estimating, say, the amount (and cost) of grass seed or turf needed to make an actual lawn (or sports field!). In 2003, a 1 kg box of grass seed sufficient for 20 square metres of grass cost about £8 to £10, and turf cost about £2.50 per square metre.

◊ From F4 onwards, pupils need to be systematic about working out missing lengths of sides. A sketch diagram is always a good idea and should be encouraged.

Safely grazing (p 85)

This investigation may be done at any suitable time during the unit.

For pupils who have become confident with decimals, 'Safely grazing' can be adapted to start from a total fence length that gives rise to non-integer sides for the pen.

Ⓐ Exploring perimeters (p 75)

A1 (a) 12 cm   (b) 12 cm   (c) 12 cm
     (d) 16 cm   (e) 18 cm

A2 (a) 40 cm   (b) 32 cm   (c) 20 cm
     (d) 22 cm

A3 Pentagon 20 cm
     Octagon 24 cm
     Hexagon 42 cm

A4 36 cm

Ⓑ Square centimetres (p 77)

B1 A 7 cm^2, B 8 cm^2, C 5 cm^2

B2 (a) A has the largest area (15 cm^2).
     (b) B has the smallest area (13 cm^2).
     (c) C and E have the same area (14 cm^2).

B3 The pupil's shapes

Ⓒ The area of a rectangle (p 78)

C1 12 cm^2; the pupil's drawings of
     1 cm by 12 cm, 2 cm by 6 cm rectangles

C2 The pupil's three rectangles such as
     1 cm by 40 cm
     2 cm by 20 cm
     4 cm by 10 cm
     5 cm by 8 cm

C3 1 cm by 16 cm
     2 cm by 8 cm
     4 cm by 4 cm (You can point out that a
     square is a special case of a rectangle.)

C4 (a) 18 cm^2 (b) 28 cm^2 (c) 25 cm^2
     (d) 12 cm^2 (e) 30 cm^2

Ⓓ Measuring to find areas (p 79)

D1 (a) 15 cm^2 (b) 24 cm^2 (c) 20 cm^2
     (d) 8 cm^2 (e) 16 cm^2 (f) 12 cm^2
     (g) 18 cm^2 (h) 21 cm^2

D2 (a) The pupil's answer (many people
     think more than half is shaded).
     (b) The area of the whole rectangle is
     42 cm^2; the area of the shaded part is
     20 cm^2. So less than half is shaded.

D3 (a) 7 cm   (b) 7 cm   (c) 4 cm
     (d) 8 cm

D4 (a) 20 cm   (b) 15 cm   (c) 60 cm
      (d) 6 cm

D5 27 cm^2

E It pays to advertise (p 81)

E1 (a) £35   (b) £48   (c) £30
      (d) £25   (e) £30   (f) £36

E2 (a) £28   (b) £64   (c) £84
      (d) £20   (e) £80   (f) £72

F Square metres (p 83)

F1 50 m^2

F2 (a) 8 m^2 (b) 4 m^2 (c) 12 m^2

F3 The pupil's sketches leading to a total area of 12 m^2

F4 (a) 14 m^2 (b) 19 m^2 (c) 15 m^2
      (d) 42 m^2 (e) 18 m^2

F5 (a) 16 m   (b) 22 m   (c) 18 m
      (d) 28 m   (e) 22 m

Safely grazing (p 85)

The biggest pen is 10 m by 10 m, with an area of 100 m^2.

What progress have you made? (p 85)

1 Area 24 cm^2, perimeter 22 cm

2 The pupil's sketch giving area 16 m^2

Practice booklet

Section A (p 23)

1 A 12 cm, B 12 cm, C 16 cm

2 (a) 38 cm   (b) 26 cm   (c) 22 cm

3 28 cm

4 60 cm

Sections B and C (p 24)

1 A 7 cm^2, B 4 cm^2, C 7 cm^2, D 2 cm^2

2 (a) 16 cm^2 (b) 18 cm^2 (c) 15 cm^2
    (d) 4 cm^2 (e) 10 cm^2

3 The pupil's three rectangles with area 24 cm^2

Sections D and E (p 25)

1 (a) 2 cm by 4 cm; 8 cm^2
   (b) 3 cm by 7 cm; 21 cm^2
   (c) 6 cm by 2 cm; 12 cm^2
   (d) 4 cm by 7 cm; 28 cm^2
   (e) 3 cm by 2 cm; 6 cm^2
   (f) 5 cm by 4 cm; 20 cm^2
   (g) 3 cm by 8 cm; 24 cm^2

Section F (p 26)

1 The pupil's sketches, giving areas
   (a) 17 m^2 (b) 14 m^2 (c) 14 m^2
   (d) 19 m^2 (e) 20 m^2 (f) 10 m^2

⑭ Rounding

T

Essential
Sheets 26 and 27
Practice booklet pages 27 to 29

Ⓐ **Place value** (p 86)

T

> Sheet 26

Place value bingo (p 86)

You need a copy of sheet 26 for yourself, and each pupil needs one of the 22 cards on the lower part of the sheet.

Read numbers in random order from the checklist at the top of the sheet, ticking them off as you go. Pupils cross numbers off with a pencil on their card.

Ⓑ **Number lines** (p 87)

> Sheet 27

Ⓒ **Nearest ten** (p 87)

A number line is used to explain rounding. It is particularly useful later as a way to demonstrate the range of values represented by a rounded value (for example, '4.7 to one decimal place' corresponds to the interval from 4.65 to 4.75).

Ⓓ **Nearest hundred, nearest thousand** (p 88)

Ⓐ Place value (p 86)

A1 (a) 5 hundreds or 500
    (b) 7 thousands or 7000
    (c) 8 tens or 80
    (d) 8 hundreds or 800
    (e) 6 tens or 60
    (f) 9 thousands or 9000

A2 (a) 4583 (b) 4673 (c) 5573
    (d) 6197 (e) 7097 (f) 6107

A3 (a) 399 (b) 6999 (c) 7090
    (d) 6900 (e) 1990 (f) 2900

A4 (a) 3482 (b) 5723 (c) 2385
    (d) 1304 (e) 4178 (f) 5062
    (g) 8390 (h) 5000 (i) 4002

Ⓑ Number lines (p 87)

B1 (a) Ten
    (b) A 210, B 350, C 490, D 560

B2 (a) Hundred
    (b) A 4200, B 5100, C 6500, D 7900

B3 (a) The pupil's arrows on the number line
    (b) The pupil's arrows on the number line
    (c) A 4250, B 4330, C 4480,
        D 5300, E 7100, F 7900,
        G 880, H 1030, I 1110, J 1220

Ⓒ Nearest ten (p 87)

C1 (a) 50 (b) 30 (c) 80
    (d) 60 (e) 60

C2 (a) 140 (b) 270 (c) 440
    (d) 860 (e) 710

C3 (a) 100 (b) 500 (c) 300
    (d) 110 (e) 390

C4 (a) 70 (b) 40 (c) 90
    (d) 100 (e) 20

C5 (a) 180 (b) 290 (c) 500
    (d) 610 (e) 360

C6 (a) 1380 (b) 1390 (c) 1400
    (d) 1400 (e) 1410

C7 (a) 4620 (b) 6390 (c) 5000
    (d) 2180 (e) 3870 (f) 2200
    (g) 3010 (h) 4740 (i) 2450
    (j) 2000

Ⓓ Nearest hundred, nearest thousand (p 88)

D1 (a) Ten (b) 6350 (c) 6400

D2 (a) 6100 (b) 6300 (c) 5900
    (d) 6000 (e) 6000 (f) 6200
    (g) 6400 (h) 6500

D3 (a) 6400 (b) 6200 (c) 6100
    (d) 6000

D4 6200

D5 (a) 6400 (b) 5900 (c) 6100
    (d) 6000 (e) 6200 (f) 6400
    (g) 6200 (h) 5900

D6 (a) 4100 (b) 7900 (c) 2500
    (d) 5100 (e) 2000 (f) 5200
    (g) 7000 (h) 4800 (i) 3500
    (j) 3000 (k) 6400 (l) 9800

D7 (a) 100 (b) 17 500 (c) 18 000

D8 (a) 17 000 (b) 18 000 (c) 17 000

D9 (a) 16 000 (b) 17 000 (c) 15 000
    (d) 16 000 (e) 19 000 (f) 18 000
    (g) 19 000 (h) 15 000 (i) 19 000
    (j) 17 000 (k) 16 000 (l) 20 000

What progress have you made? (p 90)

1 (a) 6 tens or 60   (b) 4095

2 A 420,   B 570,   C 2600,   D 3100,
  E 4400

3 (a) 40   (b) 60   (c) 60
  (d) 100

4 (a) 480   (b) 570   (c) 2480
  (d) 2700

5 (a) 6300   (b) 6500   (c) 5900
  (d) 4100

6 (a) 6900   (b) 2100   (c) 2000
  (d) 5100

7 (a) 15 000   (b) 17 000   (c) 19 000

8 (a) 18 000   (b) 23 000   (c) 40 000

Practice booklet

Sections A and B (p 27)

1 (a) 7 tens or 70   (b) 7 tens or 70
  (c) 7 thousands or 7000
  (d) 7 hundreds or 700

2 (a) 3467   (b) 3447   (c) 3557
  (d) 3357   (e) 4457   (f) 2457

3 (a) 7910   (b) 7890   (c) 8000
  (d) 7800   (e) 8900   (f) 6900

4 (a) 4137   (b) 6305   (c) 6003
  (d) 2195   (e) 4100   (f) 6109

5 (a) 420   (b) 490   (c) 610
  (d) 680   (e) 730

6 (a) 5100   (b) 5800   (c) 6900
  (d) 7700   (e) 8600

Sections C and D (p 28)

1 (a) 60   (b) 80   (c) 50
  (d) 130   (e) 260   (f) 510
  (g) 190   (h) 300

2 (a) 5710   (b) 6390   (c) 540
  (d) 7070   (e) 6450   (f) 3100
  (g) 3400   (h) 2090

3 (a) 3200   (b) 8900   (c) 5800
  (d) 3000   (e) 800   (f) 1800
  (g) 5100   (h) 4000

4 (a) 3470   (b) 4200   (c) 4000
  (d) 4200

5 2380, 2231, 2394, 2376, 2351

6 7900

7 10 700

8 (a) 18 000   (b) 17 000   (c) 14 000
  (d) 11 000

9 (a) 13 000   (b) 24 000   (c) 23 000
  (d) 21 000   (e) 35 000   (f) 24 000
  (g) 44 000   (h) 29 000

10 (a) 13 000 flee war
   (b) 46 000 visit graves
   (c) 20 000 see UFO
   (d) 65 000 at Final

11 (a) 2460   (b) 3200   (c) 7300
   (d) 7000

12 (a) 760   (b) 8000   (c) 800
   (d) 7600

⑮ Mostly multiplication

p 91	**A** Multiplying by 10, 100 and 1000	
p 92	**B** Metric units	
p 93	**C** Dividing by 10, 100, ...	
p 94	**D** The 20, 30, ... times tables	Multiplication by multiples of 10, 100, ...
p 95	**E** Area and multiplication	
p 98	**F** Tables to multiply	

Essential	Optional
	Large cards with digits 0 to 9 on them; extra 0s
Practice booklet pages 30 to 35	

Ⓐ Multiplying by 10, 100 and 1000 (p 91)

This is likely to need careful introduction. The well-known rule 'to multiply by 10 you add a 0' has limitations (it will not work for decimals). It is better to think in terms of place value.

> Optional: large (e.g. A4) cards with digits on them (0 to 9) and some extra 0s

◊ If there is enough room, you can use pupils as digits. Rule four pupil-width columns on the board (units, tens, hundreds and thousands). Give some pupils a large card each with a digit on it (0 to 9). Ask them to make a number, for example 235, against the board. Then give instructions for the group as a whole to respond to, for example:
- 'Add 40' (the pupil with 7 replaces the pupil with 3)
- 'Multiply by 10' (pupils move one place to left; pupil with 0 stands in units place)
- 'Multiply by 100' (pupils move two places to left; pupils with 0 stand in units and tens places)

◊ This activity gives an opportunity to consolidate work on place value from unit 14 'Rounding'. Encourage pupils to say the numbers aloud. It is a common misconception that the column to the left of the thousands column is millions.

◊ Many pupils will hold very strongly to the 'add a zero' strategy. It may be beneficial for some pupils to consider a few decimal examples (e.g. $1.5 \times 10 \neq 1.50$).

B **Metric units** (p 92)

You could begin the section with oral questions on conversion, based on the picture of a ruler.

C **Dividing by 10, 100, ...** (p 93)

This is also likely to need careful introduction. The well-known 'rule' that to divide by 10 you take off a zero has obvious limitations! Not all numbers end in a zero and it will not be appropriate for some that do (e.g. 3.90).

> Optional: large (e.g. A4) cards with the digits 0 to 9 on them

◊ Several activities described in section A for multiplying by 10, 100, and 1000 can be adapted to division.

D **The 20, 30, ... times tables** (p 94)

Pupils should see that multiplying by, say, 30 is equivalent to multiplying by 3 and then by 10.

In the list

The multiplications that can be made from the numbers in the list are

$20 \times 20 = 400$	$20 \times 400 = 8000$
$20 \times 40 = 800$	$40 \times 40 = 1600$
$20 \times 80 = 1600$	$40 \times 80 = 3200$
$20 \times 160 = 3200$	

This activity can easily be adapted by providing your own list or pupils could be asked to make up their own. It could also be used for divisions, of course.

E **Area and multiplication** (p 95)

◊ The teacher-led discussion should establish that a multiplication can be split into simpler multiplications and that this process can be recorded in a table. The illustrations at the beginning should provide pupils with a visual reminder of the process.

F Tables to multiply (p 98)

(p 98)

T

◊ The teacher-led discussion should establish the same points as in section E.

Three digits (p 99)

◊ With three different digits there are 6 different multiplications (counting 34 × 6 the same as 6 × 34).

34×6 43×6 36×4
63×4 46×3 64×3

The largest result is 43×6 and the smallest is 46×3.

◊ To obtain the largest result, it is clear that the smallest number must go in the units column (to minimise its effect) giving two possibilities: 43×6 or 63×4. It must be the first case as 6×3 is greater than 4×3.

In general, choose the largest digit to be the single-digit multiplier and use the next largest in the tens column.

Four digits (p 99)

◊ As there are 12 different multiplications (41×32 being the same as 32×41), it may help if pupils work in groups. After identifying the different multiplications, pupils could share the task of finding the results.

◊ For the digits 1, 2, 3, 4, the largest result is $41 \times 32 = 1312$ and the smallest is $13 \times 24 = 312$.

For the largest result, the largest numbers 4 and 3 are put in the tens place. The choice is then between 42×31 and 41×32. Think about the units × tens products. In the second case the larger units figure multiplies the larger tens figure, so this gives the larger result. A similar argument can be used for the smallest result.

Pupils may be able to give a description of the way to achieve the largest and smallest results but are unlikely to be able to give a full justification.

A Multiplying by 10, 100 and 1000 (p 91)

A1 (a) 260   (b) 2600   (c) 4800
(d) 32 400   (e) 6600

A2 (a) 300   (b) 4000   (c) 6010
(d) 30 000   (e) 80 600

A3 (a) 3000   (b) 4200   (c) 26 900
(d) 3000   (e) 4500

A4 (a) 6000   (b) 3600   (c) 84 000
(d) 2400   (e) 66 000

A5 (a) 30 000   (b) 65 000   (c) 55 000
(d) 10 000   (e) 54 000

A6 (a) $15 \times \mathbf{10} = 150$
(b) $27 \times \mathbf{100} = 2700$
(c) $\mathbf{1000} \times 34 = 34\,000$
(d) $100 \times \mathbf{24} = 2400$

A7 (a) 4060, Four thousand and sixty
(b) Twenty thousand five hundred
(c) Sixty thousand five hundred

B Metric units (p 92)

B1 (a) 60 mm   (b) 20 mm
    (c) 150 mm   (d) 250 mm
    (e) 500 mm   (f) 1300 mm
    (g) 1750 mm   (h) 2830 mm
    (i) 5000 mm   (j) 12 000 mm

B2 (a) 293 mm   (b) 288 mm   (c) 277 mm

B3 (a) 800 cm   (b) 1200 cm
    (c) 2500 cm   (d) 5000 cm
    (e) 10 000 cm   (f) 24 000 cm
    (g) 14 500 cm   (h) 37 500 cm
    (i) 80 000 cm   (j) 150 000 cm

B4 (a) 470 cm   (b) 350 cm   (c) 230 cm

B5 (a) 5000 g   (b) 15 000 g   (c) 50 000 g
    (d) 200 000 g   (e) 125 000 g

B6 (a) 1500 g   (b) 1750 g   (c) 800 g

C Dividing by 10, 100, ... (p 93)

C1 (a) 45   (b) 320   (c) 4570
    (d) 13   (e) 3600   (f) 830
    (g) 8000   (h) 102   (i) 200
    (j) 59   (k) 300   (l) 1300

C2 (a) $600 \div 10 = 60$
       $30 \div 10 = 3$
       $450 \div 10 = 45$
       $60 \div 10 = 6$
       $780 \div 10 = 78$

    (b) $600 \div 100 = 6$
       $3000 \div 100 = 30$
       $78 000 \div 100 = 780$
       $60 000 \div 100 = 600$
       $45 000 \div 100 = 450$

    (c) $3000 \div 1000 = 3$
       $78 000 \div 1000 = 78$
       $60 000 \div 1000 = 60$
       $45 000 \div 1000 = 45$

C3 (a) 19   (b) 1900
    (c) 3620   (d) 36 200 000

C4 (a) 29 cm   (b) 370 cm   (c) 50 cm
    (d) 800 cm   (e) 750 cm

C5 (a) 35 m   (b) 250 m   (c) 30 m
    (d) 3 m   (e) 10 m

D The 20, 30, ... times tables (p 94)

D1 (a) 80   (b) 120   (c) 160
    (d) 150   (e) 90   (f) 120
    (g) 200   (h) 180   (i) 300
    (j) 80

D2 (a) 600   (b) 1200   (c) 1000
    (d) 1500   (e) 2400   (f) 2000
    (g) 1800   (h) 1500   (i) 1600
    (j) 2100

D3 You could say that $40 \times 3 = 120$,
so 40×30 must be 1200.

D4 (a) 600   (b) 1000   (c) 1600
    (d) 1800   (e) 2000

D5 (a) 6000   (b) 12 000   (c) 3000
    (d) 14 000   (e) 20 000   (f) 60 000
    (g) 30 000   (h) 2100   (i) 32 000
    (j) 2500

D6 (a) 900   (b) 1600   (c) 2500
    (d) 6400   (e) 40 000

E Area and multiplication (p 95)

E1 (a) The areas are 120 cm^2 and 24 cm^2.
    (b) The large area is $120 + 24 = 144$ cm^2.

E2 (a) $350 + 28 = 378$ cm^2
    (b) $420 + 30 = 450$ cm^2

E3 (a) $180 + 12 = 192$ cm^2
    (b) $160 + 32 = 192$ cm^2
    (c) $160 + 36 = 196$ cm^2

E4

×	20	8
7	**140**	**56**

$7 \times 28 = 140 + 56 = 196$

E5

×	10	4
9	**90**	**36**

$9 \times 14 = 90 + 36 = 126$

E6 The pupil's tables leading to
   (a) $5 \times 19 = 50 + 45 = 95$
   (b) $7 \times 72 = 490 + 14 = 504$
   (c) $46 \times 6 = 240 + 36 = 276$
   (d) $83 \times 5 = 400 + 15 = 415$

E7 (a) 138 (b) 392 (c) 252 (d) 320

E8 £729

E9 £132

E10 336 days

E11 336 litres

E12

×	200	30	4
4	**800**	**120**	**16**

$$\begin{array}{r} 800 \\ 120 \\ + 16 \\ \hline 936 \end{array}$$

E13 (a) 248 (b) 309 (c) 410 (d) 450

E14 (a) 705 (b) 768 (c) 921 (d) 864

E15 (a) About 108 feet
   (b) About 744 metres
   (c) About 2667 metres
   (d) About 2889 metres
   (e) About 1456 feet

F **Tables to multiply** (p 98)

F1 (a)

×	20	5
40	800	200
7	140	35

   (b) $47 \times 25 = 800 + 200 + 140 + 35$
      $= 1175$

(c)

×	20	6
40	800	240
6	120	36

$46 \times 26 = 800 + 240 + 120 + 36$
$= 1196$

F2

×	20	4
30	600	120
7	140	28

$37 \times 24 = 600 + 120 + 140 + 28 = 888$

F3 The pupil's tables leading to
   (a) $19 \times 16 = 100 + 60 + 90 + 54 = 304$
   (b) $13 \times 27 = 200 + 70 + 60 + 21 = 351$
   (c) $21 \times 34 = 600 + 80 + 30 + 4 = 714$
   (d) $45 \times 25 = 800 + 200 + 100 + 25$
      $= 1125$
   (e) $15 \times 15 = 100 + 50 + 50 + 25 = 225$
   (f) $23 \times 91 = 1800 + 20 + 270 + 3$
      $= 2093$
   (g) $96 \times 41 = 3600 + 90 + 240 + 6$
      $= 3936$
   (h) $62 \times 87 = 4800 + 420 + 160 + 14$
      $= 5394$

F4 336 seats

F5 £851

F6 $1232 \, \text{cm}^2$

What progress have you made? (p 99)

1 (a) 40 500 (b) 3400 (c) 6000
   (d) 3 (e) 21 (f) 35

2 (a) 120 (b) 300 (c) 180
   (d) 1600 (e) 350 (f) 1200

3 (a) 282 (b) 152 (c) 6314
   (d) 252 (e) 330 (f) 1272

4 (a) 972 (b) 512 (c) 903
   (d) 2232 (e) 2214 (f) 3591

Practice booklet

Sections A and B (p 30)

1 (a) 320   (b) 3200   (c) 2560
   (d) 25 600   (e) 700   (f) 5000
   (g) 20 000   (h) 25 000

2 (a) $69 \times 10 = \mathbf{690}$
   (b) $37 \times 100 = \mathbf{3700}$
   (c) $10 \times 45 = \mathbf{450}$
   (d) $10 \times \mathbf{70} = 700$
   (e) $\mathbf{65} \times 1000 = 65\,000$
   (f) $\mathbf{50} \times 100 = 5000$
   (g) $320 \times \mathbf{10} = 3200$
   (h) $\mathbf{100} \times 90 = 9000$
   (i) $70 \times \mathbf{100} = 7000$

3 $42 \times 10 = 420$
   $42 \times 100 = 4200$
   $420 \times 10 = 4200$
   $10 \times 100 = 1000$
   $3 \times 10 = 30$
   $3 \times 1000 = 3000$
   $30 \times 100 = 3000$

4 (a) 4030; four thousand and thirty
   (b) Thirty thousand nine hundred

5 (a) 80 mm   (b) 550 mm
   (c) 1500 mm   (d) 5000 mm
   (e) 25 000 mm

6 (a) 700 cm   (b) 3500 cm
   (c) 12 500 cm   (d) 80 000 cm
   (e) 350 000 cm

7 (a) 4 kg 400 g or 4400 g
   (b) 4 kg 850 g or 4850 g
   (c) 2 kg 500 g or 2500 g

Section C (p 31)

1 (a) 22   (b) 40   (c) 8
   (d) 60   (e) 5000   (f) 30
   (g) 340   (h) 49

2 £70

3 (a) 32   (b) 80   (c) 5
   (d) 30   (e) 500   (f) 3
   (g) 450   (h) 750

4 $400 \div 100 = 4$
   $30\,000 \div 100 = 300$
   $300 \div 100 = 3$
   $4000 \div 100 = 40$
   $6500 \div 100 = 65$
   $2500 \div 100 = 25$
   $3000 \div 100 = 30$

5 $400 \div 10 = 40$
   $30\,000 \div 10 = 3000$
   $650 \div 10 = 65$
   $300 \div 10 = 30$
   $4000 \div 10 = 400$
   $250 \div 10 = 25$
   $6500 \div 10 = 650$
   $2500 \div 10 = 250$
   $30 \div 10 = 3$
   $40 \div 10 = 4$
   $3000 \div 10 = 300$

6 (a) 12   (b) 12 000   (c) 2000
   (d) 10   (e) 34 000   (f) 35 000
   (g) 206   (h) 4080

7 12 500

8 (a) 42   (b) 17 000   (c) 4500
   (d) 100   (e) 78 000   (f) 28 000
   (g) 4200   (h) 350

9 (a) 35 cm   (b) 250 cm
   (c) 75 cm   (d) 500 cm

10 (a) 45 m   (b) 72 m
   (c) 7 m   (d) 150 m

Section D (p 32)

1 (a) 1000   (b) 180   (c) 240
  (d) 2100   (e) 2000   (f) 1800
  (g) 4000   (h) 2800   (i) 2500
  (j) 36 000   (k) 9000   (l) 1800

2 30×80 and 400×6
  30×60 and 20×90
  10×20 and 5×40
  80×2 and 4×40

3 (a) $20 \times 40 = 800$
  (b) $80 \times 20 = 1600$
  (c) $80 \times 40 = 3200$
  (d) $20 \times 600 = 12\,000$
  (e) $40 \times 600 = 24\,000$
  (f) $80 \times 600 = 48\,000$

Section E (p 33)

1 (a) $360 + 12 = 372 \, \text{cm}^2$
  (b) $120 + 24 = 144 \, \text{cm}^2$
  (c) $400 + 56 = 456 \, \text{cm}^2$
  (d) $100 + 20 = 120 \, \text{cm}^2$

2

×	30	7
9	270	63

$9 \times 37 = 270 + 63 = 333$

3 The pupil's tables leading to
  (a) $7 \times 25 = 140 + 35 = 175$
  (b) $62 \times 4 = 240 + 8 = 248$
  (c) $6 \times 18 = 60 + 48 = 108$
  (d) $93 \times 5 = 450 + 15 = 465$
  (e) $8 \times 28 = 160 + 64 = 224$

4 £208

5 144

6

1	2		3		4	
4	9	6	▓	5	7	6
8	▓	5	▓	2	▓	4
▓	▓	5 1	6 8	0	▓	▓
▓	▓	▓	6	▓	▓	▓
▓	▓	7 2	4	8 4	▓	▓
9 8	▓	8	▓	3	▓	10 7
11 1	9	8	▓	12 2	7	2

Section F (p 35)

1 (a)

×	30	6
10	300	60
5	150	30

  (b) $36 \times 15 = 300 + 60 + 150 + 30 = 540$

2

×	20	8
50	1000	400
2	40	16

$28 \times 52 = 1000 + 400 + 40 + 16 = 1456$

3 The pupil's tables leading to
  (a) $24 \times 28 = 400 + 80 + 160 + 32 = 672$
  (b) $14 \times 32 = 300 + 120 + 20 + 8 = 448$
  (c) $37 \times 16 = 300 + 70 + 180 + 42 = 592$
  (d) $26 \times 54 =$
    $1000 + 300 + 80 + 24 = 1404$
  (e) $56 \times 49 =$
    $2000 + 240 + 450 + 54 = 2744$
  (f) $74 \times 12 = 700 + 40 + 140 + 8 = 888$

4 (a) $816 \, \text{cm}^2$ (b) $1161 \, \text{cm}^2$

5 £561

6 £20.40

7 The pupil's tables leading to
  (a) 7536   (b) 7614
  (c) 14 490   (d) 8784

⑯ Balancing

This work introduces, in an informal way, the idea of doing the same thing to both sides of an equation.

T p 100 **A** Scales	Introduction to the idea of 'doing the same thing' to both sides of a balance
p 100 **B** Balance pictures	Solving simple balance picture puzzles

---

Essential

OHP

Transparency copied from sheet 79

Practice booklet pages 36 and 37

Ⓐ Scales (p 100)

This section introduces the idea of scales balancing. The emphasis is on what can be done to both sides and still leave the scales in balance.

> An OHP, a transparency of sheet 79, cut up so that each picture is separate.

◊ **Picture 1**

'This was excellent with an OHP.'

'I didn't have an OHP, so I did my own silly drawings on the whiteboard.'

Use the OHP pictures of weights and rabbits to discuss with the class what you can *add* to both sides of the scales and still keep them in balance. Point out that all the weights are (notionally) 1 kilogram each.

You could also ask the pupils what you can do to the scales and definitely make them unbalanced.

Picture 2

You can then use picture 2 to discuss how to find out what one hedgehog weighs, by taking things from both sides of the scales.

Ⓑ Balance pictures (p 100)

Pupils are not expected to show any working when solving these puzzles. Of course, if they wish to write down any intermediate steps they should not be discouraged from doing so.

B Balance pictures (p 100)

B1 (a) 4   (b) 3

B2 (a) 6   (b) 2   (c) 4   (d) 5
    (e) 8   (f) 4   (g) 1   (h) 3

B3 5

B4 (a) 20   (b) 3   (c) 3   (d) 5

B5 5

B6 (a) 6   (b) 2   (c) 5   (d) 3

B7 2

B8 (a) 3   (b) 4   (c) 4   (d) 5

B9 The pupil's balance puzzle

B10 The pupil's puzzle and explanation of how to solve it

What progress have you made? (p 103)

1 4

2 The pupil's puzzle and solution

Practice booklet

Section B (p 36)

1 (a) 4   (b) 1   (c) 6

2 5

3 10

4 3

5 4

6 2

7 6

These pages are for teacher-led oral questioning, to develop mental skills with money in a 'Saturday market' context. General guidance on oral work is given in this guide for unit 8 'Oral questions: calendar'.

◊ Aim for sessions of oral questions that are regular and fairly short, with all pupils feeling a sense of achievement at the end. It is intended that you can use the pages more than once.

These sample questions are roughly in order of increasing difficulty.

1	What is the highest price marked on the stall?	£7.50
2	I have 50p. What can I afford to buy? (Invite alternatives.)	
3	What do a chocbar and a metal puzzle cost altogether?	90p
4	How much do four computer games cost?	£20.00
5	I buy a metal puzzle. How much change do I get from £1.00?	40p
6	How much do seven chews cost?	35p
7	What do a computer game and a CD cost?	£7.95
8	I buy a CD and give the stall owner £3.00. How much change do I get?	5p
9	What do a metal puzzle and a poster cost?	£1.10
10	How much do five chocbars cost?	£1.50
11	How much change do I get from a £5 note when I buy a set of Christmas tree lights?	75p
12	A woman has a £5 note. She buys a video tape. What else can she afford to buy? (Invite alternatives.)	
13	What do three cans of cola cost?	£1.35
14	How many posters can I buy for £2?	4
15	How much change do I get from a £5 note when I buy a pack of kitchen rolls?	£3.55
16	What do two video tapes cost?	£7.50
17	I buy a glue stick and a video tape. How much is that altogether?	£5.40
18	If computer games are reduced to half price how much will they be each?	£2.50
19	How much dearer is a video tape than a CD?	80p
*20	If I buy two CDs, how much change do I get from a £10 note?	£4.10
*21	Three friends buy a video tape and split the cost equally. How much does each friend pay?	£1.25
*22	I have £5. How many glue sticks can I afford?	3
*23	If I buy a T-shirt, a gent's tie, a pack of kitchen rolls and a glue stick, how much change do I get from £20?	£5.40
*24	I have £10. If I buy two sets of Christmas tree lights, how many chews can I afford to buy?	30

⑱ Gravestones

p 106 **A** What gravestones tell us	Reading from a table
p 107 **B** Making a frequency table	Grouping, tallying
p 108 **C** Comparing charts	Ways of showing frequencies
p 110 **D** Drawing a frequency chart	Follow-up questions on frequency

Essential	**Optional**
Sheet 94	Sheet 93
Practice booklet pages 38 and 39	

Ⓐ **What gravestones tell us** (p 106)

> Optional: sheet 93

◊ In some circumstances, such as a recent bereavement, sensitive handling is needed.

'A really good introduction – it captivated pupils' imagination.'

The topic obviously lends itself to locally based practical work. The sample you get from a graveyard is only of people rich enough to afford a gravestone. Sheet 93 gives a complete list of all the graves at St Mary's Church. You may find this useful for practical work.

A3 There may be discussion about which months are in winter.

Ⓑ **Making a frequency table** (p 107)

◊ Ask pupils to think of two different ways of tallying from a list.
- Go through the list, tallying all those in the 0–9 group first; then do the 10–19 group, and so on.
- Go through the list once only, tallying into the correct group as you go.

Items are less likely to be missed out if the second way is used.

◊ Pupils may already be familiar with grouping tally marks in fives: 卌

C Comparing charts (p 108)

◊ This can be done in pairs or as a class discussion.

◊ Two ways of drawing a bar chart are shown. In the first, age is treated as discrete and a gap is left between bars. In the second age is treated as continuous with no gaps between bars, so that a rule is needed for the boundaries.

D Drawing a frequency chart (p 110)

This work consolidates tallying and drawing frequency graphs.

Sheet 94

It is better if pupils work in pairs, with one reading and the other tallying. Further practice at drawing frequency graphs is in unit 37 'Graphs and charts'.

A What gravestones tell us (p 106)

A1 Probably November 1757. Remember that people are not generally buried on the day they die.

A2 Roughly 90 years

A3 It depends on what are the 'winter months'.

J	F	M	A	M	J	J	A	S	O	N	D
1		2	1		2	2	1		1	3	1

B Making a frequency table (p 107)

B1

Age (in years)	Tally	Frequency
0–9	⊞ I	6
10–19		0
20–29		0
30–39		0
40–49	I	1
50–59	III	3
60–69	III	3
70–79	I	1

The table shows that the people tended to die young or old but not in between.

C Comparing charts (p 108)

C1 There are no missing bars. They have zero height.

C2 70–80

C3 There are no people who died in these age groups.

C4 (a) Pie chart

(b) Either bar chart

(c) Either bar chart

D Drawing a frequency chart (p 110)

D1

Age group	Tally	Frequency
0–19	II	2
20–39	⊞ I	6
40–59	⊞ III	8
60–79	⊞ ⊞ ⊞ II	17
80–99	III	3

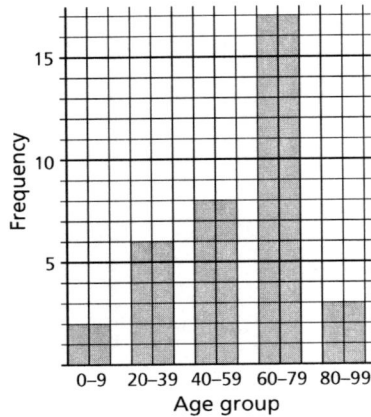

What progress have you made? (p 110)

1 (a) 12

  (b) 48

  (c) July and August

2

Estimate	Tally	Frequency
5–9	|||	3
10–14	ⷤⷮ ⷤⷮ ⷤⷮ |	16
15–19	ⷤⷮ |	6
20–24	||||	4
25–29		0
30–34	|	1

Practice booklet (p 38)

1 (a) 8   (b) 16   (c) 6

2 (a)

Age at death (years)	Tally	Frequency
0–9	ⷤⷮ ||||	9
10–19	ⷤⷮ ⷤⷮ |	11
20–29	ⷤⷮ ⷤⷮ ⷤⷮ ⷤⷮ |	21
30–39	ⷤⷮ |||	8
40–49	|||	3
50–59	||||	4
60–69	|	1
70–79	|	1
80–89		0
90–99	|	1

(b)

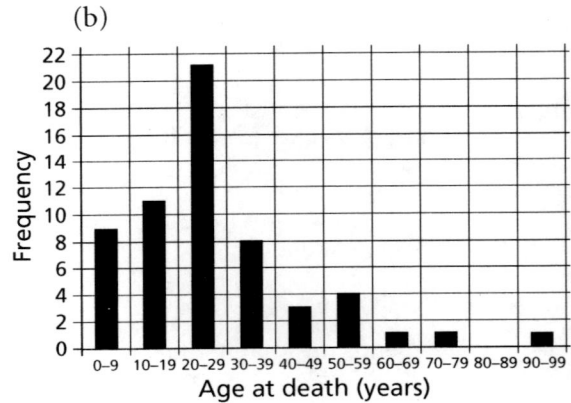

⑲ Brackets

The unit begins with a discussion of numerical expressions to help pupils see the need for brackets. It ends with 'Pam's game' which has been found motivating for pupils of all abilities. Algebraic expressions are not included.

You may have introduced brackets in earlier number work, for example, 'Four digits' from unit 1 'First bites'. This should provide consolidation and extension.

You may wish to discuss the convention that multiplication and division take priority over addition and subtraction in expressions such as $3 + 4 \times 7$ and $16 - 10 \div 2$. However, this could be confusing for some at this stage and no use is made of it in the unit.

p 111 **A** Check it out	Using brackets to indicate which part of a calculation is to be done first
	Numerical expressions that use brackets
p 113 **B** Brackets galore!	Using more than one set of brackets

> **Essential**
> Dice
>
> **Practice booklet** pages 40 and 41

Ⓐ **Check it out** (p 111)

Teacher-led discussion introduces the idea that brackets indicate which part of a calculation is to be carried out first.

◊ Some expressions are correct as they stand, for example: $21 + 3, 30 - 6$. With some expressions, brackets can be used to indicate which part of the expression needs to be evaluated first, for example (Liz's second attempt): $(3 + 3) \times 4 = 24$ but $3 + (3 \times 4) \neq 24$.
A few of the expressions are equivalent to 24 with brackets in any position, for example $(2 \times 6) \times 2$ and $5 + 3 + 20 - 4$. There is no need to labour this point at this stage.

One expression is incorrect: $13 + 21$.

◊ You can discuss how different calculators evaluate, say, $3 + 3 \times 4$, some working from left to right and others where multiplication and division take priority over addition and subtraction.

A5 Pupils can work in pairs or groups so solutions can be pooled and checked. There are six different answers but 24 calculations are possible (considering, for example, $(2 + 5) \times 3$ and $3 \times (2 + 5)$ to be different). Pupils could choose three numbers and two operations to give more or fewer than six answers.

'We had already done "Four digits" and looked at brackets and calculators. This unit was a useful revision.'

19 Brackets • 103

A6 Digits should not be joined to make larger numbers (24, 46 etc.).

As an extension, pupils could investigate expressions that give the same value with brackets in any position. For example, $(2 + 3) - 5 = 2 + (3 - 5)$ and $(2 \times 8) \div 4 = 2 \times (8 \div 4)$.

Three in a row (p 113)

'Good fun, but weaker
ones had to be
supervised closely.'

This game consolidates the use of one set of brackets.

> Three dice for each pair/group (or one dice can be thrown three times), copies of the 'Three in a row' game board (pupils can copy this onto squared paper)

◊ Pupils play the game in pairs or in larger groups (split into two teams).

◊ The squares of the game board can be made large enough to use counters.

B Brackets galore! (p 113)

Pupils work with multiple and nested sets of brackets.

◊ The activity below can be used as a homework.

> **4s make 7**
>
> Make up as many expressions as you can which have a value of 7.
>
> You can use any of $+, -, \times, \div$, brackets and 4 as often as you like.
>
> Here are some examples to start you off.
> $(44 \div 4) - 4$
> $(4 + 4) - (4 \div 4)$
> $(4 + 4 + 4) \div 4) + 4$
> $(444 + 4) - (4 \times 4 \times 4) - 44 + 4$

Pam's game (p 114)

This game consolidates the use of brackets.

> One dice if game done as a class activity, otherwise enough dice for one for each group

◊ 'Pam's game' works well as a class activity or in small groups.

In one school, a teacher split her whole class into two opposing teams and set a time limit of 1 minute. Imposing a time limit may be necessary to keep the game moving – some pupils always want to get 100 exactly!

◊ Some pupils (especially those who have completed section A only) may adopt the convention that unless brackets show otherwise, an expression is evaluated from left to right. For example, $6 \times (3 + 2) + (5 \times 4) \times 2$ may be evaluated as 100 ($6 \times 5 + 20 \times 2$ worked from left to right). Some pupils will appreciate the need for extra brackets here to give $((6 \times (3 + 2)) + (5 \times 4)) \times 2$; others may not. It is likely you will want to emphasise this point more or less strongly to different groups of pupils.

◊ You could play with fewer than six numbers or a different target.

Ⓐ **Check it out** (p 111)

A1 (a) 8   (b) 13   (c) 10   (d) 23
(e) 20   (f) 2   (g) 20   (h) 0
(i) 12   (j) 3   (k) 9   (l) 3

A2 (a) $(6 - 1) \times 3 = 15$
(b) $4 \times (1 + 2) = 12$
(c) $(2 + 1) \times 5 = 15$
(d) $(6 \div 3) + 9 = 11$
(e) $2 + (3 \times 4) = 14$
(f) $5 \times (2 - 1) = 5$
(g) $(5 - 1) \times 4 = 16$
(h) $2 + (2 \times 2) = 6$
(i) $3 \times (3 - 3) = 0$
(j) $(4 + 4) \div 4 = 2$
(k) $12 \div (3 \times 2) = 2$
(l) $10 - (6 - 2) = 6$

A3 A and Y, B and Z, C and X

A4 (a) $(1 + \mathbf{4}) \times 2 = 10$
(b) $(\mathbf{5} - 2) \times 4 = 12$
(c) $(3 \times \mathbf{3}) - 5 = 4$
(d) $2 \times (10 - \mathbf{8}) = 4$
(e) $4 \times (\mathbf{13} - 3) = 40$
(f) $(6 + \mathbf{9}) \div 3 = 5$
(g) $9 \div (\mathbf{5} + 4) = 1$
(h) $10 \div (6 - \mathbf{1}) = 2$
(i) $20 \div (\mathbf{4} + 6) = 2$
(j) $(4 \times 3) - (5 - \mathbf{2}) = 9$

A5 $(2 \times 3) + 5 = 11$
$(2 \times 5) + 3 = 13$
$(3 \times 5) + 2 = 17$
$2 \times (3 + 5) = 16$
$3 \times (2 + 5) = 21$
$5 \times (2 + 3) = 25$

There are only these six different numbers, but 24 calculations are possible.

A6 Examples are

$6 - (4 - 2)$ $6 - (4 \div 2)$
$(6 - 4) \times 2$ $2 \times (6 - 4)$
$(6 - 4) + 2$ $2 + (6 - 4)$
$(6 + 2) - 4$

Ⓑ **Brackets galore!** (p 113)

B1 $(2 + 3) \times (10 - 3) = 35$
$((4 \times 20) \div 2) \times 5 = 200$
$(10 - (2 + 5)) \times 3 = 9$
$8 + (2 \times (4 + 6)) = 28$

B2 (a) There are four different possible values:
$(5 + 3) \times (4 - 1) = 24$
$((5 + 3) \times 4) - 1 = 31$
$(5 + (3 \times 4)) - 1 = 16$ or
$5 + ((3 \times 4) - 1) = 16$
$5 + (3 \times (4 - 1)) = 14$

(b) There are five different possible values:
$(12 \div 2) + (4 \times 2) = 14$
$((12 \div 2) + 4) \times 2 = 20$
$(12 \div (2 + 4)) \times 2 = 4$
$12 \div ((2 + 4) \times 2) = 1$
$12 \div (2 + (4 \times 2)) = 1.2$

What progress have you made? (p 114)

1 (a) 15   (b) 10   (c) 10
(d) 20   (e) 2   (f) 14

2 (a) $(6 + 1) \times 2 = 14$
(b) $3 \times (10 - 8) = 6$
(c) $10 - (6 \div 2) = 7$

3 (a) 40   (b) 14

Practice booklet

Section A (p 40)

1 (a) 11   (b) 16   (c) 8
   (d) 3   (e) 4   (f) 30
   (g) 6   (h) 4   (i) 12

2 (a) $4 \times (3 + 2) = 20$
   (b) $(4 \times 3) + 2 = 14$
   (c) $10 - (3 \times 2) = 4$
   (d) $(10 - 3) \times 2 = 14$
   (e) $(4 + 5) \times 2 = 18$
   (f) $4 + (5 \times 2) = 14$
   (g) $(12 \div 3) + 1 = 5$
   (h) $12 \div (3 + 1) = 3$
   (i) $9 + (4 \div 2) = 11$

3 A and X,   B and Y,
   C and V,   D and Z

4 (a) $(2 \times \mathbf{3}) + 1 = 7$
   (b) $(3 + \mathbf{2}) \times 2 = 10$
   (c) $\mathbf{3} + (4 \times 2) = 11$
   (d) $9 - (\mathbf{3} + 2) = 4$
   (e) $(\mathbf{6} \div 2) + 5 = 8$
   (f) $12 \div (1 + \mathbf{5}) = 2$
   (g) $2 \times (8 - \mathbf{5}) = 6$
   (h) $10 - (\mathbf{7} - 3) = 6$

5 There are five different possible numbers:
$4 + (2 \times 3) = 10$ (the example)
$(4 + 2) \times 3 = 18$
$2 + (3 \times 4) = 14$
$(2 + 3) \times 4 = 20$
$3 + (4 \times 2) = 11$

These results can be found in different
ways, for example $(4 + 3) \times 2 = 14$

Section B (p 41)

1 (a) $(2 \times 6) - (3 + 1) = 8$
   (b) $2 \times (6 - (3 + 1)) = 4$

2 (a) $(12 - 5) - (2 - 1) = 6$
   (b) $((12 - 5) - 2) - 1 = 4$

3 (a) 4   (b) 10   (c) 0
   (d) 6   (e) 10

4 $((15 - 6) - 4) - 1 = 4$
$(15 - (6 - 4)) - 1 = 12$
$(15 - 6) - (4 - 1) = 6$
$15 - ((6 - 4) - 1) = 14$
$15 - (6 - (4 - 1)) = 12$

5 $((5 + 2) \times 3) + 4 = 25$
$(5 + (2 \times 3)) + 4 = 15$
$(5 + 2) \times (3 + 4) = 49$
$5 + ((2 \times 3) + 4) = 15$
$5 + (2 \times (3 + 4)) = 19$

6 (a) $(12 + 6) \div (3 - 1) = 9$
   (b) $(12 + (6 \div 3)) - 1 = 13$
   (c) $((12 + 6) \div 3) - 1 = 5$
   (d) $12 + (6 \div (3 - 1)) = 15$

Review 2 (p 115)

1 A (0, 3), B (4, 6), C (10, 3), D (4, 0)

2 (a) (7, 3) (b) (2, 3)
(c) (3, 6) (d) (4, 4)

3 (a) 20 cm (b) 21 cm^2

4 (a), (b)

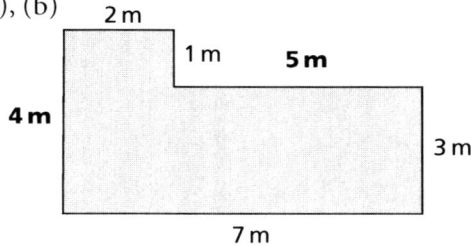

(c) The pupil's splitting up of the shape; 23 m^2

(d) 22 m

5 (a) 1790 (b) 1770 (c) 1800
(d) 1800 (e) 1810

6 (a) 5130 (b) 4670 (c) 2600
(d) 3600 (e) 1060 (f) 3750
(g) 6700 (h) 1010

7 (a) 24 900 (b) 24 800 (c) 25 100
(d) 25 000 (e) 25 200

8 (a) 3500 (b) 7600 (c) 8600
(d) 3000 (e) 2000

9 (a) 18 000 (b) 19 000 (c) 17 000
(d) 19 000 (e) 20 000

10 (a) 18 750 (b) 18 800 (c) 19 000

11 (a) 340 (b) 3400 (c) 23 400
(d) 4050 (e) 5400

12 (a) **310** × 100 = 31 000
(b) 120 × **10** = 1200
(c) **50** × 1000 = 50 000

13 (a) 600 cm (b) 3800 cm
(c) 15 000 cm (d) 90 000 cm

14 (a) 8 m (b) 12 m
(c) 20 m (d) 350 m

15 (a) 800 (b) 1800 (c) 6000
(d) 15 000 (e) 3600

16 (a) 104 (b) 852 (c) 782
(d) 2666 (e) 1206

17 (a) 1035 cm^2 (b) £504 (c) 1170 cm

18 (a) 4 (b) 14

19 (a) 4 (b) 4

20 (a) 17 (b) 10 (c) 14 (d) 20

21 (a) 2 × (5 + 3) = 16
(b) (20 − 10) ÷ 2 = 5
(c) 20 − (10 ÷ 2) = 15

22 (a) (5 + **3**) ÷ 2 = 4
(b) 12 − (**8** + 1) = 3
(c) 6 × (**6** − 1) = 30

Mixed questions 2 (p 42)

1 (a) A (0, 1), B (2, 5), C (4, 4)
(b), (c), (e)

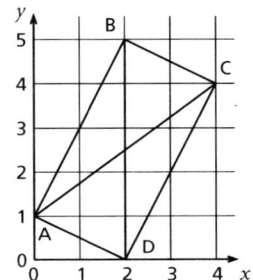

(d) Rectangle
(f) (2, 2.5)

2 (a) Perimeter 14 cm, area 10 cm^2
(b) Perimeter 16 cm, area 11 cm^2

3 (a) Perimeter 24 m, area 26 m^2

    (b) Perimeter 38 m, area 66 m^2

    (c) Perimeter 44 m, area 78 m^2

4 (a) 30   (b) 300   (c) 3000

    (d) 30

5 (a) 6500   (b) 15 000   (c) 5350

    (d) 2400

6 (a) 360   (b) 430   (c) 2100

    (d) 65 000   (e) 6000   (f) 54 000

    (g) 800   (h) 8000   (i) 52

    (j) 50   (k) 670   (l) 80

7 (a) 1500 cm   (b) 6000 cm

    (c) 2 m   (d) 15 m

    (e) 5000 g   (f) 8 kg

8 (a) 1692   (b) 1462

    (c) 3484   (d) 1656

9 (a) 4   (b) 3

10 (a) 18   (b) 8

    (c) 12   (d) 21

11 (a) 2 children   (b) 0–2 years old

⑳ Lines at right angles

For some pupils, this short unit will be a piece of revision.

p 118 **A** Thinking about right angles	Using the fact that a right angle is 90°
p 119 **B** Drawing and checking right angles	Developing accuracy

---

Essential

Set square
Plain paper

Practice booklet pages 44 and 45

A Thinking about right angles (p 118)

*'They were astonished
how many different
right angles came up.'*

Pupils may need reminding of what a right angle is, and how many degrees make up a right angle. The photographs on page 118 contrast one building consisting entirely of right angles and another where they are very hard to find. You could then ask pupils to each list 20 right angles in their classroom. Their lists can then be discussed.

A2 Some pupils may be uncertain about the points of the compass and find it difficult to relate them to turning; clockwise and anticlockwise may also be a problem. If so, it's worth developing question A2 into a class activity.

First establish the direction of north in relation to your classroom. Then have pupils take turns to stand facing in a given direction, follow instructions to turn clockwise or anticlockwise through a right angle and say what direction they are now facing.

B Drawing and checking right angles (p 119)

These simple exercises aim to develop accuracy. Pupils with poor hand–eye skills should be given sufficient time and the chance to have a second 'go' when first attempts go awry.

Set square; plain paper is essential for these questions

B5 The term 'perpendicular' is first used here. You may wish to introduce it earlier.

Ⓐ Thinking about right angles (p 118)

A1 12 and 3, 1 and 4, 2 and 5, 3 and 6,
4 and 7, 6 and 9, 7 and 10, 8 and 11,
9 and 12, 10 and 1, 11 and 2

A2 (a) North (b) North-east

(c) North-east

Ⓑ Drawing and checking right angles (p 119)

B1–B4 The pupil's drawings

B5 *a* and *b*, *c* and *j*, *d* and *e*, *i* and *l*

B6 Most people think the eight angles at the
ends of the arms are right angles but they
are not. The other four angles look too
large or too small but they are in fact
right angles.

What progress have you made? (p 121)

1 *a* and *b*, *c* and *d*, *e* and *f*

2 *a* and *e*, *b* and *h*, *c* and *f*, *d* and *g*

3 The pupil's drawing

Practice booklet

Sections A and B (p 44)

1 90 (degrees)

2 (a) 5 (b) 11

3 (a) East (b) West

(c) South-west (d) North-west

(e) South-east

4 The pupil's drawings

5 *a* and *e*, *b* and *j*, *c* and *i*, *d* and *f*, *h* and *k*.
g is the odd one out.

21 Division

Division, which involves splitting a number up into equal parts, arises in two different kinds of situation: sharing and grouping.

For example, 12 ÷ 3 can mean   '12 shared between 3'   or   'How many 3s in 12?'

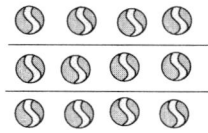

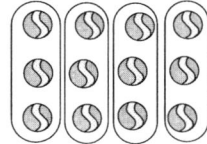

Although the answer is the same, in the first case it is the number of sweets in each group and in the second it is the number of groups.

Sharing can be done on a sharing tray (which can easily be drawn on paper). Share out one to each person, then one more each and so on.

In each round of the shareout you are taking 3 counters from the pile. This shows that sharing between 3 and grouping in 3s are equivalent.

Experience shows that pupils tend to associate division with sharing more than with grouping, so the unit starts with sharing.

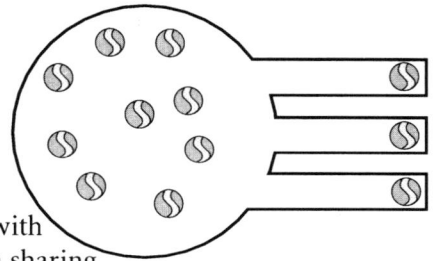

p 122 **A** Sharing	Equal sharing
p 122 **B** Divide and multiply	Linking multiplication and division
p 124 **C** Grouping	
p 125 **D** Sharing and grouping	
p 126 **E** Remainders	
p 128 **F** Division by a single-digit number: chunking	
p 130 **G** What should we do about remainders?	

Optional
Counters
Multilink
Sharing trays
Chocolate drops or Smarties
'Plates' (real or sheets of paper)

Practice booklet pages 46 to 51

21 Division • 111

Ⓐ **Sharing** (p 122)

> Optional: sharing trays, counters or multilink

◊ Start with a number of cubes or counters to represent sweets and share them out, for example 15 sweets shared between 3 people, possibly using a sharing tray. Do it by the 'one to each' process, so that every time you share out one each you are taking 3 more from the pile. This begins to show that '15 shared between 3' and '15, how many 3s?' are equivalent.

Each share is 5, so the sharing process can be written as $15 \div 3 = 5$.

◊ When you give pupils sharing problems to do practically ($12 \div 4$, for example), ask them to try to predict the result first.

Many pupils will enjoy simple division questions being given orally. Some examples are given below, but others can easily be invented.

1 8 cakes are shared equally between 2 people. How many does each person get?

2 3 girls share 12 plums equally. How many does each girl get?

3 2 boys share 14 chocolates equally. How many does each boy get?

4 10 apples are shared equally between 5 girls. How many does each girl get?

5 16 dice are shared equally between 4 boys. How many does each boy get?

6 4 girls share 8 sausage rolls equally. How many does each girl get?

7 15 buns are shared equally between 5 people. How many does each person get?

8 5 people have a meal. They share the cost equally. The meal costs £20. How much does each person pay?

Ⓑ **Divide and multiply** (p 122)

This section links multiplication facts with division facts.

> Optional: multilink

'I did this whole section orally.'

◊ One way of introducing this section is described below.

Using multilink, make 2 lots of 5 and put them together in a block.
Make 5 lots of 2 and get an identical block.

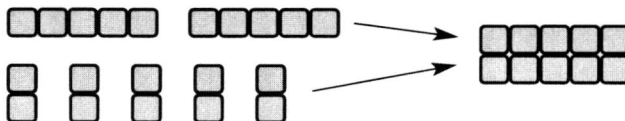

By splitting one block up into 5s and the other into 2s, you can link the four related facts:

2 × 5 = 10

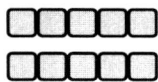

10 ÷ 2 = 5

5 × 2 = 10

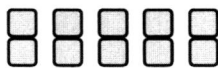

10 ÷ 5 = 2

You may need to emphasise that order is important: 2 ÷ 10 is wrong.

◊ Give pupils other combinations, such as 3 × 6, to work with (or let them think of their own). Ask them to make a complete set of multiplication and division facts in each case. Pupils could use counters or multilink to help them.

ℂ **Grouping** (p 124)

These problems all involve grouping into equal amounts.

Optional: counters or multilink

◊ 'How many boxes will you need for 15 buns packed in boxes of 3?' can be answered by considering how many 3s are in 15. Some will think of this as 3 + 3 + 3 + ... (how many times?) = 15. Others will think of **?** × 3 = 15.

Some pupils will only feel confident to tackle these problems with counters or multilink.

Encourage all pupils to relate their results to multiplication facts.

𝔻 **Sharing and grouping** (p 125)

Pupils should begin to see that sharing and grouping problems can both be solved by doing a division and that division can be thought of in terms of sharing or grouping. 80 ÷ 4 can be seen as a sharing problem (perhaps solved by halving and then halving again). However, 80 ÷ 10 is more likely to be solved as a grouping problem (how many 10s are in 80?).

Some problems are included where pupils choose whether to add, subtract, multiply or divide.

E Remainders (p 126)

Chocolate drops (p 126)

This practical activity involves sharing and remainders.

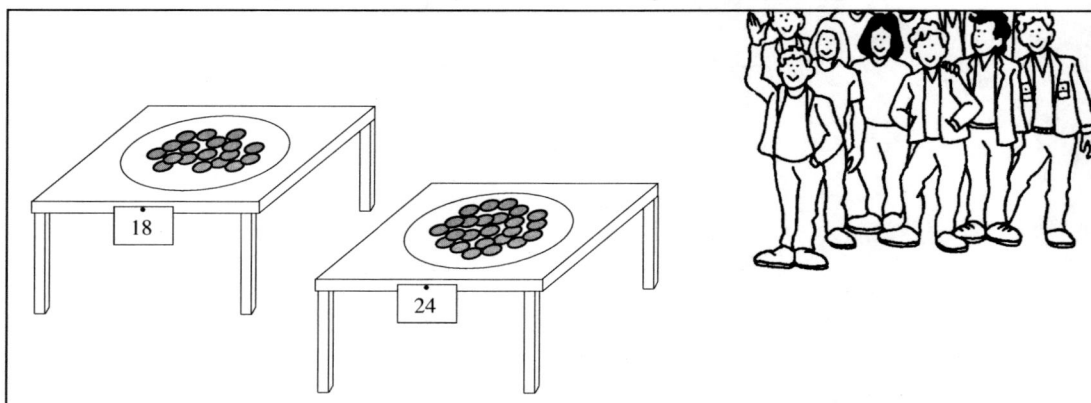

Counters, real chocolate drops or Smarties, 'plates' (real or sheets of paper)

◊ Present two 'plates' of 'chocolate drops', 24 on one and 18 on the other. Real plates and drops are likely to be more motivating, but the activity can still be done using two groups of counters or centicubes on an OHP, or by drawing on the board.

◊ Ask a pupil to choose a plate to go to.

Ask a second pupil to choose a plate. If he or she joins the first pupil they will have to share the drops on that plate equally.

Now ask a third pupil to choose, and so on. Tell the pupils that you may decide at any stage that no more pupils will be asked, and at that point the shareout will take place. (If an equal shareout is not possible, the remainder is left on the plate.)

As you go, ask pupils to explain the reasons for their choices.

You can vary the activity by starting with different numbers or having three plates (e.g. 12, 20, 24).

◊ It may help some pupils to use counters and sharing trays for the first few questions.

Sharing puzzles (p 127)

You may need to emphasise that the jewels cannot be broken in pieces.

F Division by a single-digit number: chunking (p 128)

◊ Make sure pupils appreciate the link between chunking and division.

◊ Discuss possible methods to solve problems such as 215 ÷ 6, some of which you may have covered already.

As before, encourage pupils to use the method they feel most confident with.

G What should we do about remainders? (p 130)

◊ Initially each pupil could tackle the three problems on their own. Then they can discuss their solutions in groups.

Each group could make up a problem for the whole class to do.

B Divide and multiply (p 122)

B1 (a) $8 \div 2 = \mathbf{4}$
(b) $8 \div 4 = \mathbf{2}$

B2 (a) $3 \times 2 = 6$
$6 \div 3 = \mathbf{2}$
$6 \div 2 = \mathbf{3}$
(b) $3 \times 5 = 15$
$15 \div 3 = \mathbf{5}$
$15 \div 5 = \mathbf{3}$
(c) $6 \times 2 = 12$
$12 \div 6 = \mathbf{2}$
$12 \div 2 = \mathbf{6}$

B3 (a) $3 \times 8 = 24$
$24 \div 3 = \mathbf{8}$
(b) $5 \times 5 = 25$
$25 \div 5 = \mathbf{5}$
(c) $4 \times \mathbf{6} = 24$
$24 \div 4 = \mathbf{6}$

B4 (a) 4 (b) 8 (c) 3 (d) 7
(e) 2 (f) 11 (g) 9 (h) 4

B5 (a) $28 \div \mathbf{4} = 7$ (b) $30 \div \mathbf{5} = 6$
(c) $8 \div 4 = \mathbf{2}$ (d) $\mathbf{32} \div 4 = 8$

C Grouping (p 124)

C1 6 bags   **C2** 3 bags

C3 5 boxes   **C4** 4 packs

C5 2 boxes   **C6** 7 bags

C7 7 coins   **C8** 10 days

C9 5 packs   **C10** 25 days

C11 8 boxes   **C12** 12 weeks

D Sharing and grouping (p 125)

D1 4 cakes

D2 10 packs

D3 £10

D4 (a) 7 (b) 2 (c) 12
(d) 6 (e) 8 (f) 6
(g) 5 (h) 8 (i) 8

D5 (a) 42 raffle tickets
(b) 6 taxis (c) 32 people
(d) 5 rows (e) 50 spaces

E Remainders (p 126)

E1 (a) 4 remainder 1 (b) 5 remainder 2
(c) 5 remainder 2 (d) 3 remainder 2
(e) 11 (f) 3 remainder 2
(g) 5 remainder 8 (h) 3 remainder 3
(i) 5 remainder 3 (j) 3 remainder 1

(k) 6 remainder 2   (l) 9

(m) 7   (n) 9 remainder 2

(o) 7 remainder 1

E2 (a) 2 remainder 5   (b) 4 remainder 7

(c) 4 remainder 4   (d) 3 remainder 2

(e) 2   (f) 2 remainder 2

(g) 7 remainder 2   (h) 5

(i) 3 remainder 5   (j) 7 remainder 2

E3 (a) 4 oranges   (b) 2 oranges left

E4 (a) 4 bags   (b) 1 biscuit left

E5 (a) 4 boxes   (b) 3 eggs left

E6 (a) 7 apples   (b) 1 apple left

E7 (a) 4 packs   (b) 2 oranges left

E8 3 coconuts each and 3 left over

E9 5 gums each and 2 left over

E10 (a) 9 sketchpads   (b) £1

E11 17 sweets

Sharing puzzles (p 127)

1 Each person has 9 g each (27 ÷ 3). This can be done by sharing the jewels so that one person has 2 g and 7 g, one has 3 g and 6 g and one has 4 g and 5 g.

2 The total weight is 45 g which is divisible by 1, 3, 5, 9, 15 and 45. Sharing between 9, 15 or 45 people is impossible without breaking some of the jewels. Either 3 people can have 15 g each (9 g, 6 g; 8 g, 7 g; 5 g, 4 g, 3 g, 2 g, 1 g is one way to share the jewels) or 5 people can have 9 g each (9 g; 8 g, 1 g; 7 g, 2 g; 6 g, 3 g; 5 g, 4 g).

F Division by a single-digit number: chunking (p 128)

F1 28 bags

F2 13 packs

F3 31 packs, 1 apple left over

F4 4 apples each, 5 apples left over

F5 25 boxes, 4 cakes left over

F6 24 conkers

F7 (a) 25 miles   (b) 2 rulers left over

F8 19 thimbles

F9 13 trucks

F10 26 cakes

F11 8 chocolates each, 5 remaining for the teacher

F12 The pupil's question

F13 (a) 81 tickets   (b) 128 people

(c) 26 rows   (d) 14 books

F14 The pupil's problems in words

(a) 8 rem 5   (b) 11 rem 5

(c) 26 rem 1   (d) 46 rem 2

(e) 59 rem 1   (f) 30 rem 4

G What should we do about remainders? (p 130)

G1 25 bags

G2 18 journeys

G3 29 tables

G4 17 boxes

G5 19 times

G6 The pupil's problem

G7 15 toy boxes

What progress have you made? (p 131)

1 (a) 7   (b) 6   (c) 5

2 (a) 7   (b) 4   (c) 8

3 (a) 8 rem 2   (b) 6 rem 3   (c) 7 rem 1

4 (a) 4 rem 2   (b) 4 rem 2   (c) 2 rem 2

5 (a) 4 buns, 1 bun left over

   (b) 5 boxes, 4 pens left over

6 (a) 73 rem 1   (b) 59 rem 2

   (c) 60 rem 3

7 (a) 51 rem 5   (b) 29 rem 5

   (c) 14 rem 3

8 (a) 45 taxis   (b) 17 boxes

Practice booklet

Section B (p 46)
Break the code

ARE YOU GETTING BETTER?

Section C (p 47)

1 6 bags   **2** 4 boxes

3 3 boxes   **4** 5 cards

5 5 cars   **6** 6 teams

7 7 envelopes   **8** 9 baskets

Section D (p 48)

1 5 sweets   **2** 7 packs

3 £11   **4** £9

5 5 cards   **6** 9 teams

7 58p   **8** 7 boxes

9 25 cats   **10** 6 tins

11 27 goldfish

Section E (p 49)

1 (a) Yes   (b) Yes   (c) No

   (d) Yes

2 (a) 4 rem 2 (b) 8 rem 1 (c) 4 rem 1

   (d) 6 rem 2

3 (a) 5   (b) 3 rem 3 (c) 4 rem 1

   (d) 5 rem 5

4 (a) 4 rem 7 (b) 3 rem 2 (c) 6 rem 5

   (d) 4 rem 3 (e) 2 rem 3 (f) 7 rem 1

5 (a) 6 bags   (b) 4 bangles

6 (a) 5 cars   (b) 3 tyres

7 (a) 6 lumps   (b) 2 lumps

8 8 Smarties each, with 2 left over

9 4 tea-cakes each, with 3 left over

10 6 gnomes each, with 2 left over

Section F (p 50)

1 65 boxes

2 27 tables, with 2 legs left over

3 33 stools, with 1 leg left over

4 37 packets, with 3 brushes left over

5 24 boxes, with 2 buns left over

6 29 toy cars to each shop, with none over

7 41 stamps each, with 4 left over

8 Yes, they get 114 bones each.

9 (a) 50 hats, with none left over

   (b) 33 hats, with 1 left over

   (c) 20 hats, with none left over

   (d) 16 hats, with 4 left over

10 (a) 43 rem 1   (b) 40 rem 3

   (c) 152 rem 1   (d) 28 rem 4

   (e) 30 rem 3   (f) 37 rem 2

   (g) 87 rem 2   (h) 166 rem 2

Section G (p 51)

1 7 packets

2 4 packets

3 4 boxes

4 17 times

5 Each boy gets more (95 stamps as against 86).

6 (a) 38 apples   (b) 3 more apples

7 23 times

*8 19 times

22 Parallel lines

T	p 132 **A** Looking for parallel lines	Introduction to what 'parallel' means
T	p 134 **B** Drawing parallel lines	
T	p 136 **C** Checking whether lines are parallel	

> **Essential**
> Sheets 152 and 153
> Plain paper
>
> **Practice booklet** pages 52 to 54

A Looking for parallel lines (p 132)

◊ This section introduces the term 'parallel' in a general way. Some pupils will have met the term before, so you could begin by asking pupils where they can see parallel lines in the classroom or elsewhere in the real world. You could then get a pupil to draw a pair of parallel lines on the board and ask the rest of the class to say what it is that makes the lines parallel. Clarify that parallel lines do not have to be the same length or lined up in some special way.

◊ The language of parallel lines needs to be developed and clarified carefully: 'a line parallel to *a*', 'they are parallel', 'a pair of parallel lines', 'a set of parallel lines'.

B Drawing parallel lines (p 134)

> Sheet 152, plain paper

◊ Although a method using the corner of a piece of paper is shown, pupils can use any suitable method for their patterns.

Many computer programs allow you to draw a line, copy it and drag it. Pupils can see that it remains parallel to the first line; they can also rotate it out of parallel. This may give pupils with a weak understanding of 'parallel' an important experience, and could be used to produce simple, attractive designs.

$\mathbb{C}$ Checking whether lines are parallel (p 136)

> Sheet 153

People respond to the optical illusions on sheet 153 differently, so not all pupils will be equally surprised by them. When two lines turn out not to be parallel it is sometimes worth measuring to see how much further apart they are 'at one end' compared with the other.

$\mathbb{A}$ Looking for parallel lines (p 132)

A1 (a) Line c (b) Line b

A2 (a) Line y

(b) Lines v and x are parallel.

A3 (a) True   (b) False   (c) True

A4 b and g are parallel;
c and i are parallel;
d and f are parallel.

A5 Line c with the pupil's explanation of how they decided.

$\mathbb{B}$ Drawing parallel lines (p 134)

B1 The pupil's drawings on sheet 152

$\mathbb{C}$ Checking whether lines are parallel (p 136)

C1 (a) The grey horizontal lines are parallel, though most people see the rows of tiles as tapering in alternate directions.

(b) The two black lines are not parallel! They are 7 mm further apart at the right-hand end.

C2 They are parallel.

What progress have you made? (p 136)

1 The pupil's parallel lines

2 They are not parallel.

Practice booklet

Sections A and B (p 52)

1 (a) Line p (b) Line t
(c) Lines r, u, s are parallel.

2 (a) Line h (b) Lines b, d, i
(c) Line c (d) Line g

3 m and h, g and k, c and e

4 The pupil's triangle drawing

Section C (p 54)

1 Yes, despite appearances they are parallel.

2 No

3 d is parallel to a.

㉓ Time

Essential	**Optional**
Sheet 97	Sheet 98 or 99
Scissors	OHP transparency of sheet 98 or 99
Practice booklet pages 55 to 58	

Ⓐ **Happiness graphs** (p 137)

Optional: sheet 98 or 99, and a transparency of the sheet used

◊ Discuss the 'happiness graph'. Happiness is 'measured' on a 0 to 10 scale.

◊ Pupils can then draw their own happiness graphs on squared paper or on specially ruled time graph paper (sheet 98 or 99).

'Very good. I did my own happiness graph on the board for the day I taught the class.'

Sheet 98

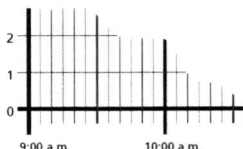

Sheet 99

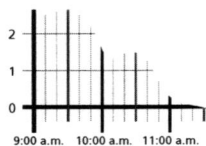

Pupils should label the axes as appropriate.

◊ Before pupils draw their graphs it may be necessary to establish the times of daily events (e.g. lesson changes, breaks). A transparency of the time graph paper is useful here.

B **Time planner** (p 137)

> Optional: sheet 98 or 99, and a transparency of the sheet used

◊ Discuss the diagram at the bottom of page 137, posing questions such as 'When does assembly end?', 'How long is break?'

◊ Discuss how you could devise a time plan for a whole week by producing a set of bars, one for each day. Pupils can draw diagrams to show their own timetables (including the weekend if they like). A shorter activity is to produce a diagram for the current day only.

If the whole group has the same timetable, each pupil could do one particular school day and then the days could be collected together to make weekly timetables.

Alternatively, each pupil could show a typical Saturday or Sunday.

As in section A, they can consider the day from 9 a.m. to 4 p.m. (sheet 98) or from 9 a.m. to 11 p.m. (sheet 99).

C **At the same time** (p 138)

> Sheet 97, scissors

◊ If you haven't already done so in sections A or B, discuss how to convert times like a quarter to seven in the morning to 6:45 a.m., and other equivalent forms.

Pupils then order the times on sheet 97. They are a set of cards that pupils can cut out and put in order. This requires quite a lot of desk space, and it may be useful for pupils to work in pairs.

The correct order is: B, L, Q, A, M, R, O, D, N, K, G, I, E, C, P, J, F, H.

◊ A simple game can be played in groups of three.
  • Shuffle the cards and deal six each.
  • Each player plays a card.
  • The latest time (or earliest, or middle, as agreed) wins the trick.
    The winner of the trick goes first in the next round.

D **Time lines** (p 138)

D9 You may need to remind pupils of the work they did on time planners in section B, where time intervals were presented as here.

E How long? (p 139)

◊ You might find it more relevant for the pupils if you discuss how to work out the lengths of some of their lessons, providing that this brings up the problem of working out an interval over two different hours.

TV listings from newspapers provide further real-life examples.

E1, 2 Pupils may find it useful to sketch a time line similar to that in the introductory example.

F The 24-hour clock (p 140)

◊ Some pupils may already be familiar with the 24-hour clock while others may not. Starting with a clock face with the 12-hour numbers, a 24-hour clock face can easily be developed. It is worth pointing out that 00:00 is midnight but 12:00 is midday. 24-hour clock times are always written with four figures (e.g. 09:40) to avoid confusion.

◊ The flight information can be used to ask a variety of questions, for example
- What time is the flight to Paris CDG in 12-hour clock time? (3:50 p.m.)
- Which flights leave after 7 p.m.? (Dubai, Aberdeen and Moscow.)
- I am catching the flight to Las Palmas.
  I arrive at the airport at a quarter past eleven in the morning.
  How long is it before my flight leaves? (2 hours 5 minutes.)
- The travel company asks me to book in at least two hours before the flight time. By what time should I book in for the flight to Dubai in a.m./p.m. time? (5:30 p.m.)

D Time lines (p 138)

D1 3:20 p.m.

D2 3:40 p.m.

D3 20 minutes

D4 4:10 p.m.

D5 30 minutes

D6 4:50 p.m.

D7 5:30 p.m.

D8 (a) 30 minutes (b) 1 hour 10 minutes

D9 1:20 p.m.

D10 30 minutes

D11 1 hour

D12 1 hour 30 minutes

D13 3:45 p.m.

E How long? (p 139)

E1 (a) 1 hour 40 minutes
    (b) 2 hours 30 minutes
    (c) 2 hours 50 minutes
    (d) 1 hour 50 minutes

E2 (a) 1 hour 45 minutes
    (b) 1 hour 40 minutes
    (c) 1 hour 45 minutes
    (d) 1 hour 30 minutes

E3 1 hour 40 minutes

E4 35 minutes

E5 15 minutes

E6 35 minutes

E7 20 minutes

E8 1:45 p.m.

E9 3:25 p.m.

E10 35 minutes

E11 (a) 1 hour 15 minutes
     (b) 1 hour 40 minutes

E12 10 minutes

E13 10:35

F The 24-hour clock (p 140)

F1 (a) 8:00 a.m.   (b) 2:30 p.m.
    (c) 10:45 a.m.   (d) 6:25 p.m.
    (e) 10:50 p.m.

F2 (a) 08:30   (b) 19:00
    (c) 11:30   (d) 21:15
    (e) 12:15   (f) 00:30

F3

12-hour clock	24-hour clock
5 p.m.	**17:00**
4:30 p.m.	16:30
11:30 a.m.	**11:30**
7:45 a.m.	07:45
3:25 p.m.	**15:25**
9:20 p.m.	21:20
10:35 p.m.	**22:35**
12:20 a.m.	00:20

F4 (a) 2:20 p.m.
    (b) 5 hours and 15 minutes
    (c) An hour and a half

What progress have you made? (p 141)

1 (a) 45 minutes
  (b) 1 hour 20 minutes
  (c) 1 hour 20 minutes
  (d) 1 hour 45 minutes

2 (a) 4:20 p.m.   (b) 9:15 a.m.
  (c) 11:10 p.m.

Practice booklet

Section C (p 55)

1 A and G, B and J, C and H,
  D and K, E and L, F and I

2 (a) 6:30 a.m.   (b) 10:15 p.m.
  (c) 8:45 p.m.   (d) 7:40 a.m.
  (e) 9:35 p.m.   (f) 9:50 a.m.
  (g) 2:55 p.m.   (h) 5:05 a.m.

Sections D and E (p 56)

1 (a) 15 minutes   (b) 25 minutes
  (c) 55 minutes   (d) 20 minutes
  (e) 1 hour 25 minutes
  (f) 1 hour 30 minutes
  (g) 1 hour 30 minutes
  (h) 6 hours 20 minutes
  (i) 2 hours 35 minutes

2 (a) 20 minutes

   (b) 1 hour 15 minutes

   (c) 20 minutes

   (d) 50 minutes

   (e) 2 hours 45 minutes

3 (a) 45 minutes

   (b) 1 hour 20 minutes

   (c) 1 hour 20 minutes

   (d) 1 hour 45 minutes

   (e) 2 hours 45 minutes

   (f) 1 hour 20 minutes

4 (a) 45 minutes   (b) 10 minutes

   (c) 1 hour 30 minutes

5 (a) 30 minutes   (b) 25 minutes

   (c) 20 minutes   (d) 45 minutes

Section F (p 58)

1

12-hour clock	24-hour clock
8 p.m.	**20:00**
9:45 a.m.	09:45
3:40 p.m.	**15:40**
5:25 p.m.	17:25
9:55 a.m.	**09:55**
11:15 p.m.	23:15

2 (a) (i) 14:00

      (ii) 16:45

      (iii) 2 hours 45 minutes

   (b) (i) 5 hours

      (ii) 8 hours 30 minutes

      (iii) 1 hour 50 minutes

   (c) 15 minutes

   (d) 7:15 a.m.

   (e) 35 minutes

24 Work to rule

The emphasis in this unit is on finding rules by analysing tile designs.
Some pupils may find the concrete experience of making the patterns
with tiles or multilink helpful.

p 142 **A** Mobiles	Finding and using a rule to calculate the number of white tiles given the number of red tiles
p 144 **B** Towers and L-shapes	Finding and using a rule that uses one operation and describing it in words
p 146 **C** Bridges	Finding and using a rule that uses one operation and describing it in words
p 147 **D** Surrounds	Finding and using a rule that uses two operations and describing it in words
p 148 **E** More designs	Investigating designs with linear rules

Practice booklet pages 59 and 60	**Optional** Tiles or multilink in two colours Squared paper

A Mobiles (p 142)

In this section pupils should become aware that there are two types of rule
for these patterns:

- by tabulating results in order we can see how the sequence of white
  tiles continues – it goes up in 2s
- by looking at the structure of the pieces we can see that the number of
  whites is equal to double the number of reds plus 3

Pupils should begin to appreciate the advantages of the latter rule.

> Optional: tiles or multilink in two colours, squared paper

◊ You can start by discussing the design of the pieces in the mobile on page
142, perhaps using tiles or multilink.

Tabulate results in order up to, say, 8 red tiles, and look at the pattern in
the table. Many pupils will spot that the number of white tiles goes up in
2s, but they should think about why it will continue in 2s.

Ask pupils to imagine the piece with, say, 100 red tiles and how we could calculate the number of white tiles. Pupils should appreciate it would take a long time to continue to add on 2s.

A discussion of the structure of the designs should lead to

- the piece with 100 red tiles has $(100 \times 2) + 3$ whites

and to the general rule

- to find the number of whites, multiply the number of reds by 2 and add 3

Use the discussion to bring out how diagrams can be useful in making the structure of the designs clear, for example:

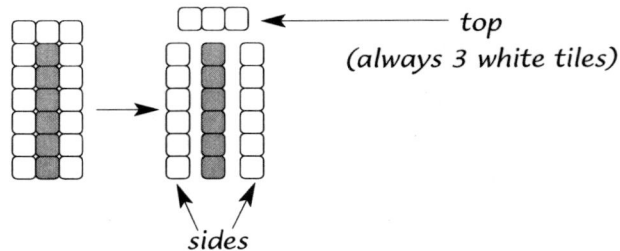

top
(always 3 white tiles)

sides

◊ When asked to explain how to find the number of white tiles for a given number of reds, some pupils may use repeated addition. Encourage all pupils to think about a rule to calculate directly the number of whites in some way. Some pupils will be able to see from the table that because the number of whites 'goes up in 2s' we must multiply the number of reds by 2 when seeking a direct rule.

A7 Pupils who give 102 white tiles as their answer may have used the rule '+ 2' to continue the sequence. Others may try to add on 2s. Encourage them to visualise the piece with 100 reds to enable them to see that the result can be found by multiplying by 2 and adding 1.

B Towers and L-shapes (p 144)

B7 Lower attainers often find it difficult to give written explanations, but they may be able to explain orally why you multiply the number of reds by 2 to get the number of whites.

C Bridges (p 146)

C7 Pupils may have counted on in 1s to answer question C6. Ask them to visualise the piece with 20 reds to help them see that the result can be found by adding 4 to the number of red tiles.

C10 Encourage pupils to draw diagrams to illustrate their explanations. Some pupils will find this difficult and should be encouraged to offer their own explanations, however tentative. An oral explanation would be perfectly acceptable at this stage.

Ⓓ **Surrounds** (p 147)

Ⓔ **More designs** (p 148)

These questions are intended to provide pupils with an opportunity to structure their own work to find rules. Encourage them to use strategies such as counting tiles in the examples given, drawing more diagrams and tabulating results to help them analyse the diagrams to find a rule.

Pupils could select one or more of these designs to investigate. They could work in groups and present their ideas to the whole class.

Ⓐ **Mobiles** (p 142)

A1 5 white tiles

A2 (a) The pupil's piece with 3 reds
(b) 7 white tiles

A3 (a) The pupil's piece with 5 reds
(b) 11 white tiles

A4

Number of red tiles	1	2	3	4	5	6
Number of white tiles	3	5	7	9	11	13

A5 (a) As the number of red tiles goes up by 1, the number of white tiles goes up by 2 each time.
(b) The pupil's explanations: for example, an increase of 1 red tile means an extra 2 white tiles, one on each side.

A6 (a) 17 white tiles (b) 21 white tiles

A7 201 white tiles

A8 The pupil's method: for example, to find the number of white tiles, multiply the number of red tiles by 2 and add 1; or a method that involves repeated addition of 2.

A9 301 white tiles

Ⓑ **Towers and L-shapes** (p 144)

B1 10 white tiles

B2 (a) The pupil's tower with 6 reds
(b) 12 white tiles

B3

Number of red tiles	1	2	3	4	5	6
Number of white tiles	2	4	6	8	10	12

B4 (a) 16 white tiles
(b) The pupil's tower with 8 reds

B5 30 white tiles

B6 80 white tiles

B7 The pupil's method: for example, multiply the number of red tiles by 2; or a method that involves repeated addition of 2.

B8 (a) 200 white tiles

(b) 116 white tiles

B9 (a) The pupil's L-shape with 5 reds

(b) 7 white tiles

B10

Number of red tiles	1	2	3	4	5	6
Number of white tiles	3	4	5	6	7	8

B11 (a) 11 white tiles

(b) The pupil's L-shape with 9 reds

B12 28 white tiles

B13 102 white tiles

B14 The pupil's method: for example, add 2 to the number of red tiles.

ℂ Bridges (p 146)

C1 10 white tiles

C2 (a) The pupil's bridge with 2 reds

(b) 6 white tiles

C3 (a) The pupil's bridge with 4 reds

(b) 8 white tiles

C4 3 red tiles

C5

Number of red tiles	1	2	3	4	5	6
Number of white tiles	5	6	7	8	9	10

C6 (a) 14 white tiles

(b) The pupil's bridge with 10 reds

C7 24 white tiles

C8 54 white tiles

C9 1004 white tiles

C10 The pupil's method: for example, add 4 to the number of red tiles.

C11 (a) 29 white tiles (b) 40 white tiles

𝔻 Surrounds (p 147)

D1 (a) The pupil's surround with 6 reds

(b) 18 white tiles

D2

Number of red tiles	1	2	3	4	5	6
Number of white tiles	8	10	12	14	16	18

D3 (a) 26 white tiles

(b) The pupil's surround with 10 reds

D4 58 white tiles

D5 206 white tiles

D6 The pupil's explanation

𝔼 More designs (p 148)

E1 (a) Set A: 102 white tiles

Set B: 202 white tiles

Set C: 401 white tiles

Set D: 204 white tiles

(b) The pupil's explanations of how to find the number of whites in each set.

What progress have you made? (p 149)

1 8 white tiles

2 (a) The pupil's drop with 5 red tiles

(b) 9 white tiles

3

Number of red tiles	1	2	3	4	5
Number of white tiles	5	6	7	8	9

4 (a) 11 white tiles

(b) 14 white tiles

5 104 white tiles

6 The pupil's explanation

7 17 white tiles

8 302 white tiles

9 The pupil's explanation

Practice booklet

1 3 whole circles

2 7 whole circles

3 The pupil's frieze; 9 whole circles

4

Number of tiles	1	2	3	4	5	6
Number of circles	1	3	5	7	9	11

5 (a) The number of circles goes up by 2 each time a tile is added.

(b) The pupil's explanation: for example, each time a tile is added one half circle on the join is completed and one circle in the centre is added.

6 (a) 15 circles   (b) 19 circles

7 199 circles

8 The pupil's explanation: for example, double the number of tiles and subtract 1.

9 299 circles

10 16 whole circles

11 The pupil's frieze; 4 circles

12 The pupil's frieze; 13 circles

13 19 circles

14

Number of tiles	1	2	3	4	5	6	7
Number of circles	1	4	7	10	13	16	19

15 (a) 28 circles   (b) The pupil's frieze

16 58 circles

17 148 circles

18 The pupil's explanation, such as 'It's the number of tiles plus twice one less than the number of tiles.'

19 (a) 73 circles   (b) 124 circles

㉕ One decimal place 7S/13

Essential

Sheet 40 and 41
Dice

Practice booklet pages 61 to 63

A Tenths (p 150)

Target (p 150)

◊ This game could be used later when you could perhaps include two decimal places.

Tenths of a centimetre (p 150)

◊ Pupils commonly view 6.4 cm as 6 cm and 4 mm, just reading the number after the decimal point as mm (so 6.04 cm would be interpreted as 6 cm and 4 mm too!). Emphasise that we can write 6 cm and 4 mm as 6.4 cm because each millimetre is one tenth of a centimetre.

◊ You could ask pupils to decide if they think 2 cm and 2.0 cm are the same or different and to justify their decision.

B Reading scales to one decimal place (p 154)

This section consolidates work from section A on the first place of decimals and includes some work on estimating. It also addresses the misconception that, when reading scales, you count the number of divisions after the whole number to give the number after the decimal point.

Dice, sheet 40 (for question B4)

Counting on (p 154)

This activity gives practice in reading scales as well as counting on in decimal steps. You could introduce this with a scale on an OHP, pointing out the steps to clarify the correct reading of the scale. More able pupils could investigate what combinations of starts and steps 'win'. Other finishing points could be used.

Counting down (p 154)

This game extends 'Counting on' to counting back.

D Ordering decimals to one decimal place (p 155)

◊ It may be helpful to draw a large number line from 0 to 8 (marked in tenths) for questions in this section.

Getting in order (p 156)

Cards made from sheet 41 (one per group)

◊ The cards include whole numbers as well as decimals, to give practice in an aspect which is often found more difficult.

◊ Another activity is to give each of, say, five pupils a sheet with a number. They have to get themselves in order as fast as they can.

A Tenths (p 150)

A1 (a) 4.9 cm   (b) 3.4 cm

A2 P 6.9 cm, Q 5.7 cm, R 8.9 cm

A3 (a) Key Q   (b) Key R

A4 In order: C 2.7 cm, B 3.5 cm, A 4 cm or 4.0 cm, D 5.1 cm

A5 a: 5 cm or 5.0 cm
    b: 5.4 cm
    b is the longer line.

A6 Red line: 6.7 cm,
    Blue line: 6.1 cm
    The red line is the longer line.

A7 Line y (3 cm) is 0.2 cm longer than line x (2.8 cm).

A8 p: 3.5 cm
    q: 3.5 cm
    r: 3.7 cm
    r is the longer line.

A9 (a) 0.9 cm   (b) 0.4 cm   (c) 0.6 cm

A10 (a) The pupil's estimate
    (b) 4.9 cm

A11 (a) 7.4 cm
    (b) 8.4 cm (longest)
    (c) 7.0 cm (shortest)
    (d) 7.7 cm

A12 A: 3.3 cm (smallest)
    B: 3.5 cm
    C: 3.7 cm (largest)

B Reading scales to one decimal place (p 154)

B1 (a) 0.9   (b) 0.6

B2 Debbie is right.

B3 (a) 8.3   (b) 9.5   (c) 10.2
    (d) 0.5   (e) 2.5   (f) 4.5

B4 A   The pupil's answers on sheet 40
     B   0.9, 3.4, 7.1, 10.6
     C   The pupil's answers on sheet 40

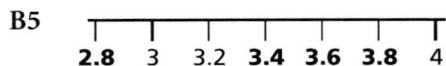

B5

2.8	3	3.2	**3.4**	**3.6**	**3.8**	4

B6 (a) 0.4   (b) 1.2   (c) 1.8
    (d) 6.6   (e) 7.4   (f) 7.8
    (g) 8.2

C Ordering decimals to one decimal place (p 155)

C1 0.8 cm, 1.7 cm, 4.3 cm, 5 cm, 5.6 cm

C2 (a) 0.7, 1.5, 2.8, 3.4, 4
    (b) 0.1, 0.4, 2, 2.9, 4.5
    (c) 0.9, 1, 6, 7, 7.6, 8
    (d) 0.7, 0.8, 1.2, 1.5, 2

C3 Kent (7 m), Jamal (6.8 m), Price (6.4 m), O'Brien (6.1 m), Stone (6 m)

C4 (a) 6 m
    (b) Davis (6.3 m), Conrad (6.2 m), Perry (6 m)

What progress have you made? (p 156)

1 About 24.5 cm (allow 24.3 to 24.7)

2 (a) 0.7   (b) 1.4   (c) 2.1

3 0.7, 1.9, 2.2, 4.3, 5

4 (a) 0.2, 0.4, 0.6, 0.8, **1.0**, **1.2**
    (b) 0.5, 0.8, 1.1, 1.4, **1.7**, **2.0**
    (c) 5, 4.7, 4.4, 4.1, **3.8**, **3.5**
    (d) 3.7, 3.2, 2.7, 2.2, **1.7**, **1.2**

Practice booklet

Section A (p 61)

1 (a) The pupil's estimates
    (b) Masked shrew   4.5 cm
       House mouse   6.4 cm
       Pygmy shrew   3.5 cm
       Harvest mouse   5.7 cm

2 (a) A 4.9 cm   B 0.9 cm   C 0.7 cm
       D 1.4 cm   E 2.5 cm
    (b) C, B, D, E, A

3 (a) The pupil's estimates
    (b) P 3.6 cm   Q 4 cm   R 4.3 cm
       S 3.8 cm   T 3.5 cm
       The longest line is 4.3 cm long (R); the shortest line is 3.5 cm (T).

Section B (p 62)

1 (a) 2.3   (b) 1.5   (c) 0.8   (d) 3.5

2 (a) 14.3   (b) 15.8   (c) 16.2
    (d) 120.5   (e) 122.5   (f) 47.7
    (g) 48.9   (h) 50.5

Section C (p 63)

1 16.7, 16.8, 17.6, 18, 18.9, 19.6, 20, 23.1

2 (a) 0.7, 1, 2.9, 3, 3.2, 4.6
    (b) 5, 5.9, 6.7, 7, 8.1, 10
    (c) 0.3, 1.1, 3.1, 3.5, 4, 5.9
    (d) 3.9, 4.8, 5, 10, 16.2, 17

3 (a) 1948 (39.9 inches)
    (b) 1960 (115.6 inches)
    (c) 1912, 1924, 1930, 1942, 1948, 1954
    (d) 1918, 1936, 1960, 1972
    (e) 1912, 1930

4 1.9, 2, 2.1

Review 3 (p 157)

1 (a) *b* and *e*, *f* and *l*, *g* and *h*
  (b) *a* and *j*, *d* and *i*, *c* and *k*

2 The pupil's drawing

3 The pupil's drawing

4 (a) 4   (b) 6   (c) 8
  (d) 7   (e) 6

5 (a) $32 \div \mathbf{4} = 8$ (b) $42 \div \mathbf{6} = 7$
  (c) $\mathbf{35} \div 5 = 7$ (d) $\mathbf{64} \div 8 = 8$

6 (a) 8 hutches   (b) 6 days
  (c) 5 boxes

7 (a) 5 rem 1   (b) 6 rem 2
  (c) 7 rem 2   (d) 8 rem 2
  (e) 6 rem 6

8 (a) 5 oranges each with 5 left over
  (b) 6 boxes with 1 cake left over
  (c) 6 bars each with 2 left to share

9 (a) 36   (b) 41   (c) 32
  (d) 31 rem 3 (e) 43 rem 2

10 (a) 17 boxes   (b) 35 tables
  (c) 16 lorries

11 (a) 8:30 a.m.   (b) 2:15 a.m.
  (c) 5:45 a.m.   (d) 9:45 p.m.
  (e) 10:40 p.m.   (f) 6:35 a.m.

12 (a) 17:15   (b) 23:30
  (c) 06:45   (d) 17:50

13 (a) 20 minutes   (b) 1 hour 45
      minutes   minutes
  (c) 25 minutes   (d) half past 12

14 (a) The pupil's drawing; 5 white tiles
  (b)

Number of red tiles	1	2	3	4	5	6	7
Number of white tiles	2	3	4	5	6	7	8

  (c) (i) 11   (ii) 51
  (d), (e) Add 1 to the number of red tiles.

15 (a) 0.8 (b) 1.3 (c) 2.6 (d) 3.5 (e) 4.6

16 (a) 0.9, 1.1, 4.9, 5, 8
  (b) 7.5, 8.7, 9.1, 9.9, 10
  (c) 0.2, 0.7, 0.8, 1, 1.1
  (d) 2.9, 3, 3.5, 5.5, 6

Mixed questions 3 (Practice booklet p 64)

1 Parallel: *a* and *e*; *b* and *j*; *c* and *h*; *d* and *k*
  Perpendicular: *f* and *l*; *g* and *i*

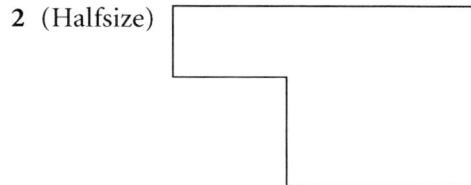

2 (Halfsize)

3 (a) 3.4   (b) 4.1   (c) 4.9   (d) 0.8
  (e) 1.6   (f) 10.2   (g) 11.4

4 (a) 0.6, 1.9, 2.1, 7
  (b) 0.1, 1.1, 1.8, 5
  (c) 0.6, 1.4, 2, 2.2

5 (a) 9, T; 4, E; 5, G; 8, R; 6, I;
     TIGER
  (b) 6, I; 4, E; 5, G; 7, L; 8, R; 3, B
     GERBIL
  (c) 6, I; 8, R; 9, T; 3, B; 2, A; 3, B
     RABBIT

6 (a) 20 boxes   (b) 5 boxes   (c) 5 days

7 (a) 20 minutes   (b) 50 minutes
  (c) 1 hour 25 minutes

8 (a)

No. of grey tiles	1	2	3	4	5
No. of white tiles	6	8	10	12	14

  (b) 24 whites   (c) 204 whites

26 Number patterns

Essential	Optional
Dice	Sheet 96

Practice booklet pages 66 to 68

A Exploring a number grid (p 160)

These investigations are all based on the six-column grid and are graded in difficulty. Some teachers have preferred to do some or all of them later in the unit.

> Optional: sheet 96

Investigation 1 (Add a number in column A to one in column B)

◊ This is a good one to start on with the whole class. It leads on to variations which pupils can investigate for themselves.

The result of A + B is always in column C (provided the grid is extended downwards). You can then ask pupils what they think could be meant by 'investigate further'. All suggestions should be responded to positively, even though they may lead nowhere.

Here are some fruitful suggestions:

What if we add two different columns?
What if we subtract?
What if we multiply?

Investigation 2 (Multiples)

◊ These occur in spatially regular patterns.

Investigation 3 (Predict the 30th number in column B, etc.)

◊ There are various ways to do this. One is to work out the 30th number in column F (30 × 6 = 180) and then work back to 176.

Investigation 4 (Predict which column 75 will be in, etc.)

◊ 75 ÷ 6 = 12 remainder 3, so 75 will be in column 3.

B **Dice numbers** (p 161)

Dice (one per pupil)

C **Magic squares** (p 162)

D **Rectangles** (p 163)

Prime numbers are introduced as numbers which cannot be made into a rectangular array, only a single line.

E **Square numbers** (p 164)

F **Missing numbers** (p 165)

Counting on you

◊ This activity could be done by writing the first three numbers on the board or an OHP, and asking a pupil to come up and write in the next number. After filling in a few numbers in the sequence, ask for an explanation of how the next number in the pattern is found. Then you could ask for a pupil to come up and write in their own first three numbers, and for others to continue the sequence.

It is intended that the sequences are linear, but pupils may suggest non-linear ways to continue.

The sequences are

4	10	16	**22**	**28**	**34**	**40**	**46**	(Add 6)
5	8	11	**14**	**17**	**20**	**23**	**26**	(Add 3)
60	54	48	**42**	**36**	**30**	**24**	**18**	(Subtract 6)
88	83	78	**73**	**68**	**63**	**58**	**53**	(Subtract 5)
100	125	150	**175**	**200**	**225**	**250**	**275**	(Add 25)
1000	950	900	**850**	**800**	**750**	**700**	**650**	(Subtract 50)

All the sequences in this section are linear.

◊ To vary the way in which tasks are presented, you could think up a sequence and write each number (for example, 6, 10, 14, 18, 22, 26, 30, 34) on a card or piece of paper without showing the pupils. Shuffle the cards, remove two and then show the rest (or read out the numbers) in no particular order. The pupils have to decide what numbers are on the missing cards. (Sometimes, if the highest or lowest card is missing, they could be left with a choice of missing numbers both of which would fit.)

Pupils could then make up their own sequence cards and do the activity with each other.

Ⓑ **Dice numbers** (p 161)

B1 (a) 2   (b) 4

(c) Top and bottom numbers add up to 7.

B2 (a) Can see 7, can't see 14

(b) Can see 12, can't see 9

(c) Can see 10, can't see 11

(d) Can see 6, can't see 15

B3 (a) Yes (b) 6, 7, 9, 10, 12, 14, 15

(c) 15 (d) No

B4 (a) 27 (b) 26 (c) 26 (d) 25

Ⓒ **Magic squares** (p 162)

C1 15

C2 (a)

7	2	9
8	6	4
3	10	5

(b)

6	11	4
5	7	9
10	3	8

(c)

10	5	6
3	7	11
8	9	4

C3 (a)

7	14	9
12	10	8
11	6	13

(b) The new square is a magic square. The magic number is 30 because there is an extra 15 in each row, etc.

C4 (a)

16	3	2	13
5	10	11	8
9	6	7	12
4	15	14	1

(b)

15	10	3	6
4	5	16	9
14	11	2	7
1	8	13	12

Ⓓ **Rectangles** (p 163)

D1 (a) The pupil's 3 by 8 rectangle

(b) 3 × 8 (or 8 × 3). Don't count a single line (1 × 24) as a rectangle.

D2 10×2

D3 (a) 2×6, 3×4

(b) 2×8, 4×4

(c) 2×9, 3×6

(d) 2×15, 3×10, 5×6

D4 You can only make a single line with 17.
Other numbers include 2, 3, 5, 7, ...

D5 11, 13, 17

D6 Because even numbers can be made into
a rectangle $2 \times$ something

D7 (a) 11   (b) 17   (c) 19   (d) 23

(e) 29   (f) 31   (g) 37

E Square numbers (p 164)

E1 $6 \times 6 = 36$ and $7 \times 7 = 49$

E2 64, 81, 100

E3 3, 5, 7, 9, ... odd numbers

E4 (a) 16   (b) 25   (c) 64   (d) 121

E5 Rob has done 10×2. He should have
done 10×10.

E6 (a) 400   (b) 900

(c) 2500   (d) 6400

E7 (a) 13   (b) 33   (c) 73

(d) 61   (e) 155

*E8 $6 = 2^2 + 1^2 + 1^2$

$7 = 2^2 + 1^2 + 1^2 + 1^2$

$8 = 2^2 + 2^2$

$9 = 3^2$

$10 = 3^2 + 1^2$

$11 = 3^2 + 1^2 + 1^2$

$12 = 2^2 + 2^2 + 2^2$

$13 = 3^2 + 2^2$

$14 = 3^2 + 2^2 + 1^2$

$15 = 3^2 + 2^2 + 1^2 + 1^2$

$16 = 4^2$

$17 = 4^2 + 1^2$

$18 = 3^2 + 3^2$

$19 = 3^2 + 3^2 + 1^2$

$20 = 4^2 + 2^2$

$21 = 4^2 + 2^2 + 1^2$

$22 = 3^2 + 3^2 + 2^2$

$23 = 3^2 + 3^2 + 2^2 + 1^2$

$24 = 4^2 + 2^2 + 2^2$

$25 = 5^2$

$26 = 5^2 + 1^2$

$27 = 5^2 + 1^2 + 1^2$ or $3^2 + 3^2 + 3^2$

$28 = 5^2 + 1^2 + 1^2 + 1^2$ or
$\qquad 4^2 + 2^2 + 2^2 + 2^2$ or
$\qquad 3^2 + 3^2 + 3^2 + 1^2$

$29 = 5^2 + 2^2$

$30 = 5^2 + 2^2 + 1^2$

F Missing numbers (p 165)

F1 (a) 17, 20   (b) add 3

F2 (a) 36, 43   add 7

(b) 11, 7   subtract 4

(c) 56, 67   add 11

(d) 28, 21   subtract 7

F3 (a) 2, 8, 14, **20**, 26, **32**, 38

(b) 5, 9, **13**, **17**, 21, **25**, 29

(c) **2**, 11, 20, **29**, 38, 47, **56**

(d) 36, 31, **26**, **21**, 16, 11, **6**

F4 15

F5 (a) **14**, **17**, 20, 23, **26**, **29**, 32

(b) **54**, **46**, 38, 30, **22**, 14, **6**

F6 29, 47

F7 (a) 7, **10**, 13, **16**, 19, **22**, 25

(b) 1, **5**, 9, **13**, **17**, 21, **25**

What progress have you made? (p 166)

1 Second row × 2 is in first row.
First row × 2 is in second row.
Third row × 2 is in third row.

2 (a) The pupil's 2 × 5 rectangle
 (b) 2 × 5

3 Two of 2 × 18, 3 × 12, 4 × 9, 6 × 6

4 (a) 13 can only be divided by 1 and
 by 13. It can't be arranged in a
 rectangle, only a line.
 (b) 23, 29

5 (a) 25 is 5 × 5   (b) 16   (c) 36

6 (a) 81   (b) 144   (c) 1600

7 (a) 5, 9, 13, 17, 21, **25**, **29**   add 4
 (b) 43, 37, 31, 25, 19, **13**, **7**   subtract 6
 (c) **10**, **17**, 24, 31, 38, 45, **52**   add 7

Practice booklet

Sections B and C (p 66)

1 3

2 12

3 (a) 10   (b) 15   (c) 14

4 (a) 23   (b) 32   (c) 24

5
1	14	4	15
8	11	5	10
13	2	16	3
12	7	9	6

Sections D and E (p 67)

1 (a) Rectangle pattern 3 × 6
 (b) Rectangle pattern 2 × 16

2 (a) 2 × 4
 (b) 2 × 14, 4 × 7
 (c) 2 × 6, 3 × 4
 (d) 2 × 18, 3 × 12, 4 × 9, 6 × 6

3 No, 9 makes a square 3 × 3.

4 (a) 23   (b) 29   (c) 7

5 (a) 29, 31   (b) 41   (c) 47

6 (a) 9   (b) 49   (c) 81   (d) 100

7 (a) 1600   (b) 4900   (c) 8100

8 5^2 obviously makes a square. Prime
numbers don't make rectangles
(or squares).

9 (a) 7   (b) 16   (c) 37   (d) 36

10 16, 1, 9 and 25

Section F (p 68)

1 (a) 22 and 26
 (b) You add 4 each time.

2 (a) 25, 28   add 3
 (b) 30, 26   take off 4
 (c) 80, 93   add 13
 (d) 45, 34   take off 11

3 (a) 28, 40   (b) 31, 34, 40
 (c) 14, 32, 50   (d) 44, 41, 32

4 17

5 (a) 38, 26   (b) 16, 46

㉗ Rectangles

Pupils consider properties of rectangles and learn about translations.
Earlier topics such as parallel lines, perpendicular lines, coordinates and
reflection symmetry are revised.

p 167 **A** Properties

p 168 **B** Using coordinates

> **Essential**
> Centimetre squared paper
> **Practice booklet** page 69

Ⓐ **Properties** (p 167)

On the grid (p 167)

> Centimetre squared paper

You may wish to discuss whether a 2 by 1 rectangle is the same as a 1 by 2
rectangle. The possible sizes on a 3 cm by 3 cm grid are:

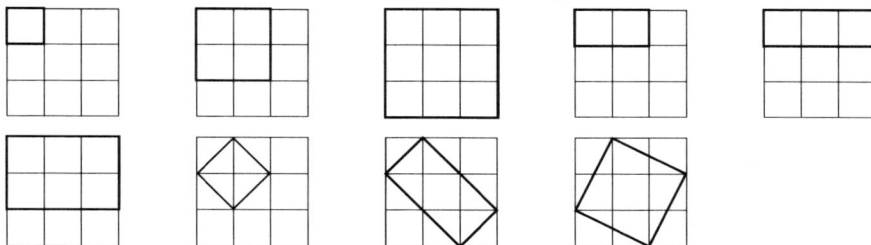

As an extension pupils could consider the different positions on the grid
that each size of rectangle can occupy.

A6 Many pupils may think that the diagonals are lines of symmetry; you may
need to demonstrate that they are not with a mirror.

◊ At the end of section A you could summarise the properties of a rectangle:
  • Opposite sides are parallel and of equal length.
  • The angles are all right angles, so adjacent sides are perpendicular.
  • A rectangle, unless it is a square, has only two lines of symmetry.
  • The diagonals are of equal length.

Ⓑ **Using coordinates** (p 168)

> Squared paper

Ⓐ Properties (p 167)

A1 (a) 6 cm   (b) 3 cm

A2 A right angle

A3 (a) 6.7 cm   (b) 6.7 cm

A4 (a) Side AB is **perpendicular** to side BC.

    (b) Side AD is **parallel** to side BC.

    (c) Side AB is **parallel** to side CD.

A5 The pupil's rectangle PQRS where PQ is 4 cm and QR is 5 cm

A6 (a) False   (b) True   (c) True

    (d) False   (e) True

Ⓑ Using coordinates (p 168)

B1 (a) E $(1, 3)$ (b) F $(5, 3)$

    (c) G $(5, 1)$ (d) H $(1, 1)$

B2, B3 (a)

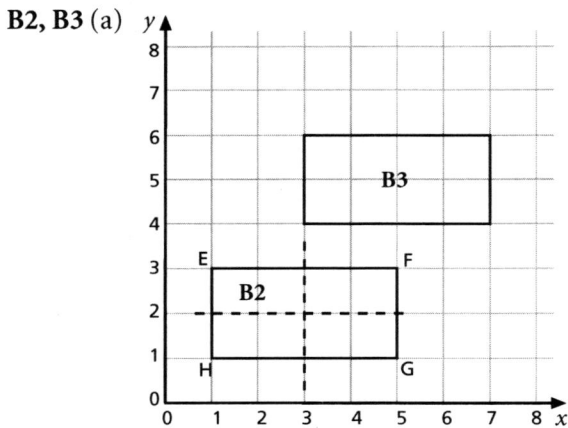

B3 (b) E′ $(3, 6)$, F′ $(7, 6)$, G′ $(7, 4)$, H′ $(3, 4)$

B4

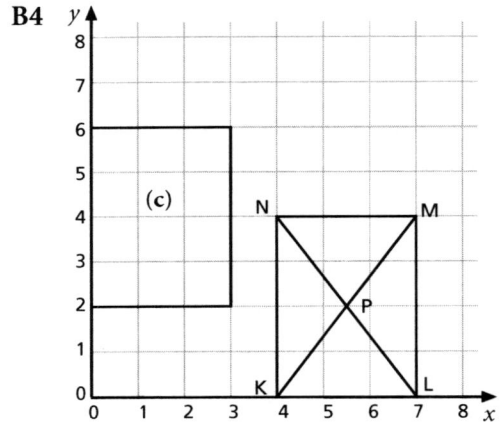

  (b) Point P has coordinates $(5\frac{1}{2}, 2)$.

  (c) K′ $(0, 2)$, L′ $(3, 2)$, M′ $(3, 6)$, N′ $(0, 6)$

What progress have you made? (p 168)

1 (a) Two, unless it is a square

   (b) Right angles

2 E′ $(0, 7)$, F′ $(4, 7)$, G′ $(4, 5)$, H′ $(0, 5)$

Practice booklet

Sections A and B (p 69)

1 (a) True   (b) False   (c) True

   (d) True   (e) False

2 (a) $(1, 6)$ (b) $(3, 6)$ (c) $(3, 3)$

   (d) $(1, 3)$

3, 4

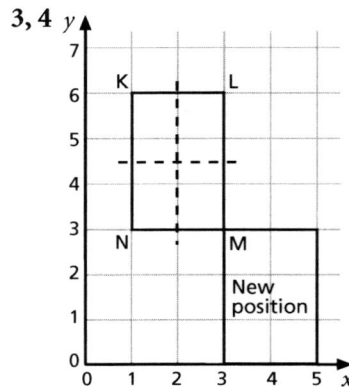

5 $(5, 0)$

28 Oral questions: measures (p 169)

The information table in the pupil's book is the basis for oral questions

General advice on oral work is given in the notes for unit 8, 'Oral questions: calendar'.

◊ Start by explaining what is meant by the diameter of a ball, and that the diameter may vary slightly, which is why a range of values is given (for example, the men's shot can be any measurement between 11 and 13 cm in diameter). Similarly a range is given for the weight and cost in some cases. Encourage pupils to include units such as grams, centimetres and £.

These sample questions are roughly in order of increasing difficulty.

1	What is the diameter of a golf ball?	4.3 cm
2	What is the diameter of a pool ball?	5.7 cm
3	What is the diameter of a tenpin bowl?	21.6 cm
4	How much is a rounders ball?	£4
5	How much is a golf ball?	£1.30
6	Which is larger – a table tennis or golf ball?	Golf ball
7	What is the weight of a golf ball?	45.9 g
8	What is the weight of a table tennis ball?	2.5 g
9	What is the weight of a tenpin bowl?	7258 g
10	What is the cost of a woman's shot?	£10
11	What is the cost of two men's shots?	£20
12	How much do three rounders balls cost?	£12
13	How much is the cheapest football?	£7.50
14	Which is the smallest ball?	Table tennis
15	Which is the heaviest ball?	Men's shot
16	Which balls could weigh 160 grams?	Cricket, pool

The numbers have one decimal place.

Optional
Sheet 125
Large cards with digits on them

Practice booklet pages 70 to 73

Ⓐ **Adding and subtracting** (p 170)

Both mental and written methods are included, but the emphasis is on mental methods.

◊ Draw a scale on the board marked in tenths from 0 up to 3.
Use this as a number line to show ways of adding and subtracting. For example, to do 1.7 + 0.5 you can start at 1.7, add on 0.3 to get to 2 and that leaves an extra 0.2 still to add. You can then show the process on a line on which only the whole numbers are marked, like this:

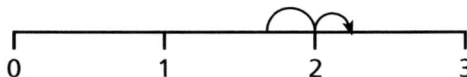

The reason for removing the tenths marks is to force a mental method rather than counting the marks.

◊ The pictures of the fish and the animals are for oral questions, such as
  • How much does the big/small fish weigh?
  • How much do the two fish weigh altogether?
  • How much more does the big fish weigh than the small one?
  • How much more does the cat weigh than the guinea pig?

◊ Some pupils may find it helpful to replace, say, 5 by 5.0 when doing written calculations with one decimal place.

One or two (p 170)

Optional: sheet 125

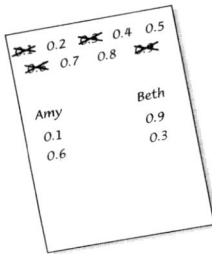

The game can be played without the cards. Players write down the nine numbers from 0.1 to 0.9 on a single piece of paper. As they choose a number, they cross it out and write it on their side of the paper.

The game can also be played by two people in front of the class, writing the numbers on the board.

B **Problems** (p 172)

C **Multiplying by a whole number** (p 173)

◊ A common error is to fail to take over 10 tenths into the units column (e.g. 1.4 × 3 given as 3.12). The diagram in the pupil's book shows that 3 lots of '4 tenths' = 12 tenths = 1 unit and 2 tenths. You may need to do other examples like this, using the number line.

Having dealt with examples of the type 0.3 × 4, you can go on to the type 2.3 × 4.

D **Rounding** (p 174)

A number line approach is used, similar to that in unit 14 'Rounding'.

E **Estimating by rounding** (p 174)

◊ It is worth discussing estimation with pupils. Many think of it as an inferior process to calculation, and do not see that it may lead them to correct a false answer obtained with a calculator.

F Place value (p 176)

T

| Optional: large cards with digits on them |

◊ Rule columns on the board: hundreds, tens, units, tenths. Give a number, for example 267.4. Pupils holding digits make the number by standing in the right columns. You can then ask them to make a new number by, say, adding 0.1 to the number they have already. (For example, the 'tenths' pupil holding 4 would be replaced by another holding 5.)

◊ The same set-up can be used to show multiplication and division by 10 (see section G).

G Multiplying and dividing by 10 (p 177)

If pupils have learned a rule 'to multiply by 10, add a nought', they will find it breaks down where decimals are involved.

| Optional: large cards with digits on them |

T

◊ Before, or instead of, the calculator-based activities, you could use 'pupil digits' as described in the previous section.

When a number (e.g. 62.3) has been made, you can ask for it to be multiplied by 10. Pupils should know that the 60 becomes 600 and the 2 becomes 20, so you can focus attention on what happens to the 0.3.

Dividing by 10 can be approached in a similar way, as the inverse of multiplying by 10.

A Adding and subtracting (p 170)

A1 (a) 1.1 (b) 1.1 (c) 1.7
(d) 1.9 (e) 4.5

A2 (a) 2.9 (b) 5.9 (c) 8.8
(d) 6.7 (e) 6.8

A3 (a) 3.4 (b) 3.1 (c) 3.6
(d) 3.8 (e) 5.5

A4 (a) 0.3 (b) 1.7 (c) 0.3
(d) 4.5 (e) 2.9

A5 (a) 1.5 (b) 3.7 (c) 1.3
(d) 1.3 (e) 0.6

A6 (a) 0.3 (b) 1.3 (c) 3.6
(d) 2.8 (e) 3.8

A7 (a) 3.4 (b) 0.4 (c) 3.5
(d) 7.2 (e) 7.2

A8 (a) Mel (b) Sam

A9 (a) 13.4 kg (b) 9.2 kg (c) 1.6 kg
(d) 2.6 kg (e) 4.2 kg

A10 (a) 12.8 (b) 10.3 (c) 8.2
(d) 11.5 (e) 14.4

A11 (a) 3.3 (b) 4.5 (c) 5.4
(d) 5.3 (e) 5.8

A12 (a) 2.9 (b) 6(.0) (c) 10.3
(d) 3.8 (e) 15.5

B Problems (p 172)

B1 1.7 cm

B2 11.6 km

B3 4.7 m

B4 0.6 m

B5 1.8 litres

B6 3.3 km

B7 6.4 kg

B8 The pupil's problems, one 'add', the other 'subtract'

C Multiplying by a whole number (p 173)

C1 (a) 0.8   (b) 1.2   (c) 1.5
     (d) 1.6   (e) 2.0 (or 2) (f) 2.4

C2 7.5 litres

C3 (a) 9.6   (b) 30.1   (c) 16.8
     (d) 4.2   (e) 21.6   (f) 29.6

C4 (a) 4.2 kg   (b) 13 kg or 13.0 kg
     (c) 20.4 kg   (d) 22.4 kg

C5 20.8 metres

*C6 $28.4 \times 7 = 198.8$
The truck can carry all the boxes.

D Rounding (p 174)

D1 4

D2 (a) 7   (b) 5   (c) 12
     (d) 16   (e) 20   (f) 40
     (g) 8   (h) 1   (i) 120
     (j) 300

D3 (a) 4   (b) 18   (c) 20
     (d) 32   (e) 150

D4 (a) 3   (b) 4   (c) 6
     (d) 8   (e) 2   (f) 13
     (g) 26   (h) 11   (i) 29
     (j) 60

E Estimating by rounding (p 174)

E1 (a) Estimate $4 + 1 + 5 = 10$, actual 9.7
     (b) Estimate $6 + 2 + 1 = 9$, actual 9.2
     (c) Estimate $2 + 4 + 6 = 12$, actual 11.4
     (d) Estimate $4 + 7 + 1 = 12$, actual 12.3

E2 (a) Estimate $3 + 4 + 9 + 6 = 22$, actual 22
     (b) Estimate $2 + 5 + 6 + 8 = 21$, actual 20.8

E3 (a) Estimate $4 + 3 + 7 + 1 = 15$, actual 14.8
     (b) Estimate $8 + 6 + 1 + 6 = 21$, actual 20.6

E4 An estimate is $10 + 4 + 5 + 4 + 1 = 24$, which is a lot more than 18.8.

E5 (a) Estimate $14 - 6 = 8$, actual 8.1
     (b) Estimate $18 - 8 = 10$, actual 10.3
     (c) Estimate $12 - 7 = 5$, actual 5.5
     (d) Estimate $19 - 11 = 8$, actual 8.1

E6 (a) Estimate $14 - 7 = 7$, actual 7.7
     (b) Estimate $17 - 14 = 3$, actual 3.2
     (c) Estimate $14 - 3 = 11$, actual 11.4

E7 (a) Estimate $3 \times 3 = 9$, actual 8.7
     (b) Estimate $6 \times 2 = 12$, actual 11.8
     (c) Estimate $7 \times 4 = 28$, actual 27.2

E8 (a) Estimate $4 \times 6 = 24$, actual 22.8
     (b) Estimate $6 \times 5 = 30$, actual 30.5
     (c) Estimate $4 \times 6 = 24$, actual 25.2

E9 (a) Roughly 6×2 m $= 12$ m
     (b) 11.4 m

F Place value (p 176)

F1 (a) 1 thousand (1000)
     (b) 6 hundreds (600)
     (c) 4 tens (40)
     (d) 5 units (5)

F2 (a) 4 tens (40)   (b) 9 tenths (0.9)
    (c) 2 hundreds (200)

F3 (a) 46.6   (b) 55.6   (c) 45.7
    (d) 135.8   (e) 136.7   (f) 145.7

F4 (a) 33.8   (b) 42.8   (c) 32.9
    (d) 147.4   (e) 137.5   (f) 138.4

F5 50

F6 7

F7 0.7

F8 (a) 6.7   (b) 1.3   (c) 2.1
    (d) 10.9   (e) 0.6

F9 (a) $2\frac{9}{10}$ (b) $7\frac{3}{10}$ (c) $8\frac{1}{10}$ (d) $\frac{3}{10}$

Ⓖ **Multiplying and dividing by 10** (p 177)

G1 (a) 34   (b) 2   (c) 14
    (d) 21   (e) 3   (f) 17
    (g) 54   (h) 151   (i) 27
    (j) 204

G2 (a) 137   (b) 532   (c) 18
    (d) 46   (e) 56   (f) 33
    (g) 7   (h) 194   (i) 325
    (j) 160

G3 (a) 4.3   (b) 2.7   (c) 13.3
    (d) 64.8   (e) 0.6   (f) 6.7
    (g) 2.5   (h) 0.4   (i) 4.2
    (j) 12.5

G4 (a) 3.6   (b) 0.9   (c) 1.6
    (d) 7 or 7.0   (e) 3.1   (f) 13.6
    (g) 5.5   (h) 13   (i) 2.1
    (j) 0.7

G5 (a) 47   (b) 3.6   (c) 380
    (d) 1   (e) 26.8   (f) 15.7
    (g) 386   (h) 7.4   (i) 710
    (j) 1.9

What progress have you made? (p 178)

1 (a) 1.2   (b) 2.4   (c) 5.2
   (d) 3.5   (e) 4.8   (f) 4.5

2 (a) 11.5   (b) 3.8   (c) 9.2
   (d) 0.7   (e) 10.5   (f) 1.2

3 (a) 3 or 3.0   (b) 10.8

4 (a) Estimate 2 + 6 + 9 = 17, actual 16.3
   (b) Estimate 15 − 13 = 2, actual 1.8
   (c) Estimate 6 × 7 = 42, actual 40.6
   (d) Estimate 10 + 3 + 8 = 21, actual 20.8

5 (a) 7 tens (70)   (b) 3 tenths (0.3)

6 (a) 56   (b) 36.2   (c) 23.1
   (d) 43   (e) 0.8   (f) 7

Practice booklet

Sections A and B (p 70)

1 (a) 10.6   (b) 7.4   (c) 6.6
   (d) 12.3

2 (a) 1.4   (b) 0.9   (c) 1.8
   (d) 4.5

3 (a) 7.9   (b) 7(.0)   (c) 0.6
   (d) 4.7

In questions 4–7 the 'odd answer out' is in bold.

4 (a) 9.1   (b) **8.9**   (c) 9.1

5 (a) **12.6**   (b) 12.2   (c) 12.2

6 (a) 6.6   (b) 6.6   (c) **6.5**

7 (a) **3.2**   (b) 3.3   (c) 3.3

8 (a) 12.4 cm   (b) 2.8 cm

9 (a) 13.8 km   (b) 2.6 km

10 (a) Andy 4.3 kg, Fi 4.5 kg
    (b) Fi, by 0.2 kg

Section C (p 71)

1 6 kg

2 4.2 kg

3 (a) 2.4   (b) 3 or 3.0   (c) 4 or 4.0
   (d) 3.5   (e) 0.4   (f) 1.5
   (g) 1.8   (h) 2.1

4 (a) 6.8   (b) 12.6   (c) 11.2
   (d) 36.4   (e) 5.4   (f) 51.2
   (g) 5.4   (h) 56.7

5 20.4 metres

6 6.3 metres

7 Yes, the lift can carry the boxes.
   $46.2 \times 8 = 369.6$
   369.6 kg is less than 370 kg so it is just possible.

Sections D and E (p 72)

1 (a) 5   (b) 15   (c) 20
   (d) 6

2 (a) 40   (b) 60   (c) 200
   (d) 2000

3 (a) 2300   (b) 7700   (c) 100
   (d) 400

4 (a) 2680   (b) 8900   (c) 401
   (d) 6300

5 (a) $6 + 2 + 8 + 4 = 20$
   (b) 20.2

6 (a) Estimate $4 + 2 + 8 = 14$; 14.2
   (b) Estimate $8 + 4 + 1 = 13$; 12.3
   (c) Estimate $17 - 4 = 13$; 13.6
   (d) Estimate $24 - 10 = 14$; 14.4
   (e) Estimate $8 + 5 + 7 + 5 = 25$; 24.5
   (f) Estimate $3 + 8 + 5 + 1 = 17$; 17.9
   (g) Estimate $4 \times 4 = 16$; 15.6
   (h) Estimate $2 \times 6 = 12$; 10.8
   (i) Estimate $6 \times 3 = 18$; 17.7
   (j) Estimate $8 \times 4 = 32$; 31.2

Sections F and G (p 73)

1 (a) 5 tens (50)
   (b) 8 tenths (0.8)
   (c) 6 hundreds (600)

2 (a) 33.7   (b) 32.8   (c) 42.7
   (d) 106.5   (e) 116.4   (f) 107.4

3 (a) 48.6   (b) 57.6   (c) 47.7
   (d) 284.5   (e) 275.5   (f) 274.6

4 (a) 3.3   (b) 4.7   (c) 6.1
   (d) 8.9

5 (a) $1\frac{7}{10}$ (b) $\frac{9}{10}$ (c) $6\frac{3}{10}$
   (d) $2\frac{1}{10}$

6 (a) 68   (b) 6   (c) 18
   (d) 315   (e) 45   (f) 2
   (g) 206   (h) 11

7 (a) 7.5   (b) 1.6   (c) 0.8
   (d) 14.7   (e) 0.4   (f) 2
   (g) 0.3   (h) 9.9

8 (a) 490   (b) 4.9   (c) 60.3
   (d) 106   (e) 52.1   (f) 3.2
   (g) 27   (h) 19

This unit introduces probability through games of chance. Probability is
expressed as a fraction in situations with equally likely outcomes.

p 179 **A** Chance or skill?	Deciding whether games are skill or chance
p 180 **B** Fair or unfair?	Deciding whether a game of chance is fair
p 181 **C** Probability	The probability scale from 0 to 1
p 182 **D** Spinners	Writing a probability as a fraction
	Finding the probability of an event not happening

Essential	**Optional**
Dice, counters of different colours	OHP transparencies of sheets 111 to 115
Sheets 111 to 115	
Practice booklet pages 74 and 75	

A **Chance or skill?** (p 179)

> Dice and counters
> Sheets 111 to 113 (game boards)

◊ Before discussing and playing the games, you could get pupils talking about
chance, for example, the National Lottery. People often have peculiar ideas
about chance. For example, would they write on a National Lottery ticket
the same combination as the one that won last week? If not, why not?

You could ask pupils to think about games that they know and to discuss
the elements of chance and skill in them.

◊ You can copy the games on to card and laminate them or cover them with
clear adhesive film. The games can then be used more than once.

◊ Before playing each game, ask pupils to try to decide from its rules whether
it is a game of pure chance, a game of skill, or a mixture.

Some games of skill give an advantage to the first player. Who goes first is
usually decided by a process of chance.

◊ You could split the class into pairs or small groups, with each group playing
one of the games and reporting on it.

*'Some pupils
thought that
using a dice
meant it was
all chance.'*

◊ 'Fours' is a game of skill. 'Line of three' is a mixture of chance and skill. 'Jumping the line' appears to involve skill, because you have to decide which counters to move and it looks as if you can get 'nearer' to winning. But it is a game of pure chance. At any stage there is only one number which will enable the player to win. If any other number comes up, whatever the player does leaves the opponent in essentially the same position.

⒝ Fair or unfair? (p 180)

Pupils have to decide whether a game of chance is fair or unfair.

> Dice, counters, sheets 114 and 115
> Optional: transparencies of sheets 114 and 115

◊ You can start by playing the game several times as a class, with a track on the board. You may need to go over the meaning of 'odd' and 'even'.

◊ When pupils play the game themselves, ask them to record the results and then pool the class's results.

◊ Let pupils consider each other's ways of making the game fairer (if they can think of any!). Do they agree that they would be fairer?

'I split the class into groups and gave rat numbers to each group. There was intense rivalry!'

◊ For the first rat race, counters ('rats') are lined up at the start. The teacher rolls a dice. The score tells which rat moves forward one square. Everyone chooses a rat they think will win. For the second rat race, two dice are rolled and the total is used.
In the first race, some pupils may believe that 6 is 'harder' to get than other numbers. If so, you could discuss this.
In the second race you could ask for suggestions for making it fairer, still using two dice. (For example, the track could be shortened for the 'end' numbers – even so, Rat 1 is never going to win!)

⒞ Probability (p 181)

The probability scale from 0 to 1 is introduced.

◊ Construct a probability 'washing line' by pinning the ends of a long piece of string to the board. Mark 0 and 1 at the ends of the line. Pupils can then be asked to hang cards for different events on the line with paper clips. Explain first the meanings of the two endpoints of the scale. Something with probability 0 is often described as 'impossible'. However, there are different ways of being impossible and some of them have nothing to do with probability (for example, it is impossible for a triangle to have four sides). So it is better to use 'no chance of happening' or 'never happens'. Something with probability 1 always happens, or is certain to happen.

◊ Go through the events listed in the pupil's book and discuss where they go on the scale. The coin example leads to the other especially important point on the scale, $\frac{1}{2}$. Associate this with 'equally likely to happen or not happen', with fairness, 'even chances', etc.

◊ Keep the approach informal. The important thing is to locate a point on the right side of $\frac{1}{2}$, or close to one of the ends when appropriate (for example, in the case of the National Lottery!).

D **Spinners** (p 182)

Pupils write probabilities as fractions.

◊ A spinner is very useful in connection with probability. It shows fractions in a familiar way. This is a good opportunity to revise and reinforce basic fraction work. Emphasise that the parts (sectors) into which the spinner is divided have to be equal.

D3 If pupils give $\frac{1}{3}$ for (d), they have ignored the inequality of the parts.

Make your own spinner (p 184)

This is a good opportunity to do some further tallying and bar charts, as well as accurate drawing. The shape of the bar chart will show how fair a spinner is.

C **Probability** (p 181)

C1 (a) Unlikely   (b) A 50% chance
     (c) Impossible   (d) Certain
     (e) Likely

C2

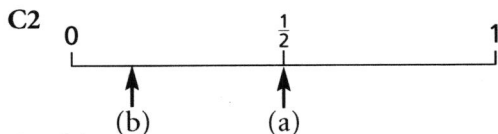

C3 (a) A boy
     (b)

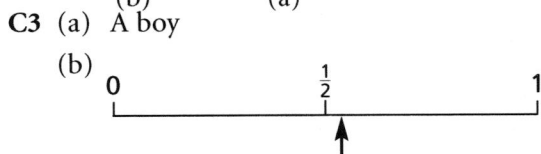

C4 (a) P   (b) R   (c) Q

D **Spinners** (p 182)

D1 (a) $\frac{1}{2}$ (b) $\frac{1}{4}$ (c) $\frac{1}{6}$ (d) $\frac{1}{8}$

D2 $\frac{2}{5}$

D3 (a) $\frac{2}{6}$ or $\frac{1}{3}$ (b) $\frac{3}{8}$ (c) $\frac{5}{8}$ (d) $\frac{1}{4}$

D4 (a) $\frac{1}{6}$ (b) $\frac{2}{6}$ or $\frac{1}{3}$ (c) $\frac{3}{6}$ or $\frac{1}{2}$

D5 (a) $\frac{1}{8}$ (b) $\frac{5}{8}$ (c) $\frac{2}{8}$ or $\frac{1}{4}$

D6 $\frac{4}{5}$

D7 (a) $\frac{5}{6}$ (b) $\frac{5}{8}$ (c) $\frac{3}{8}$

D8 (a) $\frac{1}{4}$ (b) $\frac{3}{4}$ (c) $\frac{3}{8}$

What progress have you made? (p 184)

1 The game is unfair.
B wins more often than A.

2 (a) L   (b) M   (c) K

3 (a) $\frac{1}{5}$ (b) $\frac{4}{5}$ (c) $\frac{2}{5}$ (d) $\frac{3}{5}$

Practice booklet

Section C (p 74)

1 (a) Unlikely   (b) Likely
   (c) A 50% chance   (d) Impossible

2

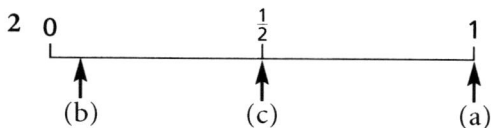

3 (a) Q   (b) P   (c) R

Section D (p 75)

1 (a) $\frac{3}{10}$ (b) $\frac{2}{10}$ or $\frac{1}{5}$
   (c) $\frac{5}{10}$ or $\frac{1}{2}$ (d) $\frac{5}{10}$ or $\frac{1}{2}$

2 (a) $\frac{3}{8}$ (b) $\frac{4}{7}$ (c) $\frac{1}{6}$ (d) $\frac{2}{9}$

3 (a) $\frac{5}{8}$ (b) $\frac{3}{7}$ (c) $\frac{5}{6}$ (d) $\frac{7}{9}$

㉛ Negative numbers

> **Essential**
> Sheets 90 to 92
>
> **Practice booklet** pages 76 to 78

Ⓐ Colder and colder (p 185)

> Sheets 90 to 92

◊ You could start by asking pupils for examples of temperatures which they think they know. Can they give a reasonable estimate of the room temperature? Have they seen different types of thermometer? Where?

◊ You could ask pupils to mark their suggestions for the temperatures A, B, C, … on a number line. Some pupils may have little idea of some of them.

'We had quite a heated (no pun intended) discussion on these and they were amazingly accurate in their final list.'

Temperature trumps (p 188)

This can be played in groups of 2, 3 or 4.
All the cards are dealt, face down. Players do not look at them.
The player whose turn it is reveals their top card and chooses either summer temperature or winter temperature.
The others turn over their top card.
If summer temperature was chosen then the highest summer temperature wins. If winter temperature was chosen then the lowest winter temperature wins.
The winner takes the turned-over cards and has the next turn.
The overall winner is the player with most cards at the end.

'Superb game! Worked well with all abilities.'

Ⓑ Temperature changes (p 189)

◊ The bulleted questions in the introduction are examples of the types that could be asked about the map. Other maps and data from newspapers or websites would provide further written or oral practice.

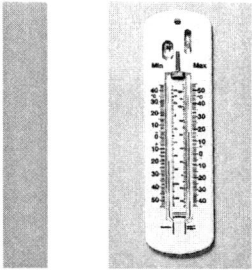

◊ A maximum/minimum or freezer thermometer would be a useful aid here. A max/min thermometer has markers which show the highest and lowest temperature occurring since it was last set.

ℂ **Using tables** (p 190)

𝔻 **Temperature graphs** (p 191)

Pupils extract information from graphs.

◊ Information on maximum and minimum temperatures from a school weather station, regional weather centre, internet source or local enthusiast would make a useful stimulus for discussion.

𝔸 **Colder and colder** (p 185)

The approximate temperatures in order are

J   Oven temperature for baking a cake 200°C

I   Temperature of a hot bowl of soup 55°C

G   Temperature of a hot bath 40°C

B   Human body temperature 37°C

H   Temperature at which butter melts 35°C

C   Temperature of a hot summer's day in Britain 30°C

L   Temperature of a heated swimming pool 25°C

E   Temperature inside an ordinary fridge 5°C

K   Temperature of ice-cream when it's good to eat 2°C

F   Temperature inside a car in the morning after a frosty night 1°C

D   Antarctic sea water temperature ⁻1°C

A   A winter's day temperature at the north pole ⁻30°C

A1

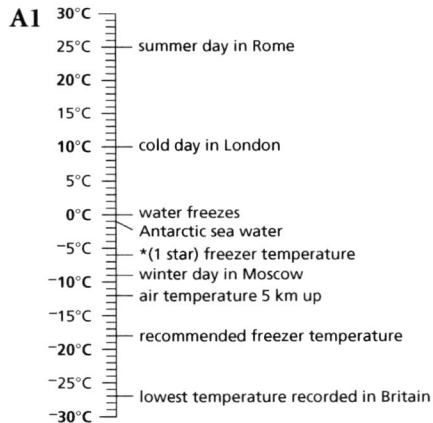

A2 (a) ⁻8°C (b) ⁻2°C (c) 2°C
(d) ⁻9°C

A3 (a) Glasgow 1°C   Paris ⁻10°C
London ⁻3°C   Oslo 0°C
Moscow ⁻5°C

(b) Paris   (c) Glasgow

A4 ⁻4°C

A5 ⁻12°C

A6 ⁻15°C, ⁻5°C, 0°C, 3°C, 7°C

A7 (a) 3°C < 9°C   (b) ⁻4°C < 1°C

(c) ⁻1°C < 3°C   (d) ⁻2°C > ⁻5°C

(e) ⁻5°C < 3°C   (f) ⁻3°C < ⁻1°C

(g) 4°C > ⁻5°C   (h) 0°C > ⁻3°C

A8 (a) P 5°C, Q ⁻5°C, R ⁻25°C, S ⁻5°C

(b) R

A9 (a) Freezer   (b) Fridge

(c) Neither   (d) Fridge

A10 (a) Yes   (b) No   (c) No

(d) Yes   (e) No   (f) Yes

B Temperature changes (p 189)

B1 (a) 3°C   (b) ⁻1°C   (c) ⁻4°C

(d) ⁻4°C   (e) ⁻8°C   (f) 2°C

B2 13 degrees

B3 15 degrees

B4 Twelve true sentences are possible.

15°C is	5 degrees higher than	10°C
	10	5°C
	15	0°C
10°C is	5 degrees higher than	5°C
	10	0°C
	15	⁻5°C
5°C is	5 degrees higher than	0°C
	10	⁻5°C
	15	⁻10°C
0°C is	5 degrees higher than	⁻5°C
	10	⁻10°C
⁻5°C is	5 degrees higher than	⁻10°C

B5 Twelve true sentences are possible.

⁻10°C is	5 degrees lower than	⁻5°C
	10	0°C
	15	5°C
⁻5°C is	5 degrees lower than	0°C
	10	5°C
	15	10°C
0°C is	5 degrees lower than	5°C
	10	10°C
	15	15°C
5°C is	5 degrees lower than	10°C
	10	15°C
10°C is	5 degrees lower than	15°C

C Using tables (p 190)

C1 (a) 7 days   (b) 13 degrees

C2 (a) True   (b) False

(c) True   (d) True

C3 (a) Scott   (b) Nord

(c) 10 degrees   (d) 9 months

(e) June, July, August

(f) November, December, January

D Temperature graphs (p 191)

D1 (a) 7°C   (b) 8 p.m.

(c) ⁻28°C   (d) Midnight

(e) 7°C   (f) 16

D2 (a) ⁻2°C   (b) ⁻13°C

(c) 11:30 p.m. and 1 a.m.

(d) $2\frac{1}{2}$ hours

D3 (a) January and February

(b) June, July and August

(c) 9 degrees

D4 (a) 25°C   (b) ⁻5°C

(c) January and February

(d) 8 degrees   (e) 7 degrees

(f) October (12 degrees)

What progress have you made? (p 193)

1 ⁻32°C, ⁻14°C, ⁻3°C, ⁻1°C, 2°C, 15°C

2 (a) 23 degrees   (b) 5 degrees
   (c) 8°C

3 (a) ⁻3°C   (b) 6 degrees
   (c) Between 9 p.m. and 11 p.m.
   (d) 2°C

Practice booklet

Section A (p 76)

1 A, D, C, B, E

2 (a) R (b) U (c) Q (d) S (e) P

3 (a) ⁻12°C   (b) ⁻4°C

4 (a) 0°C   (b) 1°C

5 (a) 8°C > 5°C   (b) 4°C < 12°C
   (c) ⁻2°C < 5°C   (d) ⁻7°C < ⁻5°C
   (e) 4°C > ⁻6°C   (f) ⁻3°C > ⁻5°C

6 ⁻4°C, 3°C, ⁻1°C, 5°C

Sections B and C (p 77)

1 (a) 8°C   (b) ⁻8°C

2 5°C

3 ⁻10°C

4 6°C

5 ⁻9°C

6 ⁻36°C

7 A temperature of **5**°C is **20** degrees higher than a temperature of ⁻**15**°C.

8 47 degrees

9 (a) January and December
   (b) July   (c) 20 degrees

Section D (p 78)

1 7°C

2 ⁻2°C

3 Wednesday

4 Saturday

5 (a) 5 degrees   (b) 7 degrees
   (c) 7 degrees

6 Wednesday (9 degrees)

32 Action and result puzzles (p 194)

In each puzzle, the action cards show operations to be performed on a starting number and the result cards show the results. Pupils match up the results with the actions.

The puzzles provide number practice and an opportunity to apply some logical thinking. They may reveal misconceptions about number.

Essential	Optional
Puzzles on sheets 102 to 104	Sheet 110 (blank cards)
Scissors	OHP transparencies of some sheets, cut into puzzle cards
Practice booklet pages 79 and 80	

T

The puzzles are listed below, roughly in order of difficulty.

Sheet 102   6 puzzle (+ and –, whole numbers ≤ 10)
31 puzzle (+ and –, whole numbers ≤ 50)
4 puzzle (× and ÷ by 1, 2, 4 and 5)

Sheet 103   5 puzzle (+, –, × and ÷, whole numbers ≤ 20)
16 puzzle (+, –, × and ÷, whole numbers ≤ 32)
270 puzzle (+ and –, multiples of 10)

Sheet 104   44 puzzle (+ and –, whole numbers ≤ 100)
60 puzzle (+, –, × and ÷, simple two- and three-digit numbers)
5.5 puzzle (+ and –, decimals with 0.5 only)

'I copied the cards on to pieces of acetate which could be moved about on the OHP. Pupils went to the OHP to show how the cards matched up.'

◊ This has worked well with pupils sitting in pairs on tables of four. When each pair has matched the cards, all four pupils discuss what they have done. An aim is to encourage mental number work. However, pupils may want to do some calculations and demonstrate things to their group using pencil and paper. It is not intended that a calculator should be used.

◊ Puzzles that pupils find easy can be done without cutting out the cards: they simply key each action card to its result card by marking both with the same letter. However, something may be learnt from moving cards around to try ideas out before reaching a final pairing.

◊ Solutions can be recorded by
  • keying cards to one another with letters as described above
  • sticking pairs of cards on sheets or in exercise books
  • writing appropriate statements, such as 8 – 3 = 5

◊ After pupils have solved some puzzles, they can make up some of their own (using the blank cards) to try on a partner. This may tell you something about the limits of the mathematics they feel confident with.

Sheet 102

6 puzzle

Action	Result
− 2	4
+ 3	9
− 4	2
+ 4	10
+ 2	8
− 3	3

31 puzzle

Action	Result
+ 10	41
− 10	21
+ 9	40
− 9	22
− 19	12
+ 19	50
− 7	24
+ 15	46

4 puzzle

Action	Result
× 2	8
÷ 2	2
× 4	16
÷ 4	1
÷ 1	4
× 5	20

Sheet 103

5 puzzle

Action	Result
+ 10	15
× 4	20
− 3	2
+ 7	12
× 2	10
× 3	15
+ 5	10
÷ 1	5

16 puzzle

Action	Result
÷ 4	4
− 4	12
+ 4	20
÷ 8	2
− 8	8
× 2	32
÷ 2	8
+ 5	21

270 puzzle

Action	Result
+ 100	370
− 80	190
+ 300	570
− 60	210
+ 30	300
− 150	120
+ 240	510
− 100	170

Sheet 104

44 puzzle

Action	Result
− 43	1
+ 43	87
− 29	15
+ 29	73
+ 50	94
− 9	35
+ 56	100
− 14	30

60 puzzle

Action	Result
÷ 10	6
− 10	50
− 12	48
÷ 5	12
× 3	180
× 10	600
+ 100	160
÷ 3	20

5.5 puzzle

Action	Result
+ 0.5	6
− 0.5	5
+ 2.5	8
− 2.5	3
− 1.5	4
+ 9	14.5
+ 3.5	9
− 3.5	2

Practice booklet (p 79)

1 15 puzzle

Action	Result
+ 3	18
+ 7	22
− 7	8
− 3	12
− 10	5
+ 5	20
− 5	10
+ 10	25

2 (a) 6   (b) 24

3 (a) add 2   (b) add 6

   (c) subtract 11

4 20 puzzle

Action	Result
× 2	40
+ 11	31
− 7	13
− 13	7
÷ 2	10
÷ 5	4
× 5	100
+ 30	50

5 (a) 11   (b) 30   (c) 5   (d) 120

6 (a) subtract 5   (b) add 13
   (c) add 60 *or* multiply by 4

7 add 40, *and* multiply by 3

Review 4 (p 195)

1 (a) The pupil's pattern for 36
  (b) 2×18, 3×12, 4×9, 6×6
  (c) Yes, $36 = 6^2$

2 (a) 7   (b) 14   (c) 36

3 (a) 1600   (b) 4900   (c) 1100

4 (a) None of them is prime.
  (b) 4, 36 and 49 are square.

5 (a) 19, 22. The rule is add 3.
  (b) 64, 58. The rule is subtract 6.
  (c) 3.2, 3.7. The rule is add 0.5.
  (d) 65, 74. The rule is add 9.
  (e) 8.2, 7.9. The rule is subtract 0.3.

6 (a) Rectangle
  (b) $12\,cm^2$ (c) 14 cm
  (d), (e)

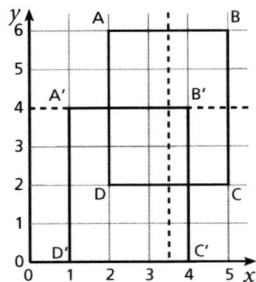

  (f) A (1, 4) B (4, 4) C (4, 0) D (1, 0)

7 (a) 3.5 (b) 3.4 (c) 4.1 (d) 5.8
  (e) 1.7 (f) 1.2 (g) 3.8 (h) 4.6
  (i) 4.8 (j) 4.5

8 (a) 2.6 (b) 4.8 (c) 3.6 (d) 3.2
  (e) 10.4 (f) 19 (g) 25 (h) 150
  (i) 273 (j) 183 (k) 5.5 (l) 0.9
  (m) 10 (n) 1.7 (o) 6.5

9 0.9 metre

10 (a) 4.8 kg   (b) 6.8 kg

11 (a) 9 kg
  (b) He is carrying 1.4 kg less than Cora.

12 (a) 13   (b) 8   (c) 10
  (d) 111   (e) 10

13 (a) $4 + 3 + 7 = 14$; 13.9
  (b) $13 − 8 = 5$; 4.9
  (c) $1 \times 6 = 6$; 5.4
  (d) $4 + 4 + 13 = 21$; 20.9
  (e) $20 − 13 = 7$; 7.1
  (f) $2 \times 8 = 16$; 16.8

14 (a) Unlikely   (b) A 50% chance
  (c) Certain

15

0 ———————————————— 1
  ↑(c)   ↑(a)   ↑(b) ↑(d)

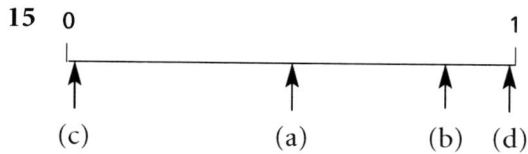

16 (a) $\frac{2}{4}$ or $\frac{1}{2}$ (b) $\frac{2}{3}$
   (c) $\frac{4}{6}$ or $\frac{2}{3}$ (d) $\frac{2}{5}$

17 (a) $\frac{2}{4}$ or $\frac{1}{2}$ (b) $\frac{1}{3}$
   (c) $\frac{2}{6}$ or $\frac{1}{3}$ (d) $\frac{3}{5}$

18 (a) Cracow   (b) Cadiz
   (c) Oslo by 3 degrees
   (d) 7 degrees colder
   (e) Oslo and Cadiz
   (f) 4°C

19 ⁻18°C, ⁻10°C, ⁻1°C, 0°C, 2°C, 12°C

20 (a) 2°C < 6°C   (b) 2°C > ⁻6°C
   (c) ⁻2°C < 6°C   (d) ⁻2°C > ⁻6°C

Mixed questions 4 (Practice booklet p 81)

1 (a) The pupil's rectangle with 30 dots, either 3 by 10 or 2 by 15
   (b) A pair of factors of 30
   (c) No; you cannot make a square with 30 dots.

2 (a) 81   (b) 28   (c) 3600

3 (a) 5, 9, 13, 17, 21, **25**, **29**;   add 4
   (b) 85, 79, 73, 67, **61**, **55**;   subtract 6
   (c) 34, 39, **44**, 49, 54, **59**;   add 5
   (d) 45, **41**, 37, **33**, 29, **25**;   subtract 4

4 (a) 5.4 kg   (b) 1.8 kg
   (c) 3.4 kg   (d) 14.4 kg

5 (a) Estimate 4 + 2 + 6 = 12; actual 11.7
   (b) Estimate 19 − 8 = 11;   actual 10.8
   (c) Estimate 4 × 5 = 20;   actual 18.5

6 (a) 46.5   (b) 27.7
   (c) 137.3   (d) 254.4

7 (a) 38   (b) 246
   (c) 26.5   (d) 3.9

8 (a) Q   (b) P   (c) R

9 (a) $\frac{1}{8}$ (b) $\frac{4}{8} = \frac{1}{2}$
   (c) $\frac{3}{8}$ (d) $\frac{5}{8}$

10 (a) Living room
   (b) Garden
   (c) 7 degrees
   (d) The patio; 6 degrees
   (e) 3°C

11 ⁻7°C, ⁻4°C, ⁻2°C, 0°C, 8°C, 10°C

12 (a) 5°C   8°C
   (b) 7°C   ⁻5°C
   (c) ⁻4°C   ⁻7°C

33 Two decimal places

7S/13

This unit extends decimal notation to two decimal places.

Essential
Sheets 223 and 224
Metre rules

Optional
Tape measure marked in decimals of a metre
OHP transparency of sheet 222
Large cards with digits 0 to 9 on them

Practice booklet pages 83 to 87

A Tenths review (p 198)

B Tenths and hundredths (p 199)

Optional: OHP transparency of sheet 222

T

◊ The diagram on page 199 shows a litre measuring cylinder marked in tenths and hundredths. You can ask why the new subdivisions are called 'hundredths', and relate this to the fact that there are 100 of them in the whole unit.

◊ It is useful to have one of these 'measuring cylinders' copied on to an OHP transparency, using sheet 222. By cutting out the shaded strip, or using a coloured piece of acetate, you can then vary the height of the shaded column and ask for the measurement in decimal form.

◊ Emphasise the place values: tenths, hundredths.

◊ Numbers below 0.1 are dealt with in section C, but you may wish to introduce them here.

C **Less than 0.1** (p 201)

> Sheet 223 for question C3
> Optional: OHP transparency of sheet 222

◊ The OHT of sheet 222 is useful again here.

◊ Pupils often find it more difficult to recognise decimals like 0.02 when reading a scale. This may be because in other instances the tenths mark below shows how the decimal begins. (For example, 0.42 is above 0.4.)

D **0.3 and 0.30 are equal** (p 202)

'I emphasised the usefulness of using decimals with the same number of decimal places, in particular for addition, subtraction and writing in order.'

This is another aspect of decimals that confuses pupils, made worse by their tendency to read, for example, 1.30 as 'one point thirty' (like 'one pound thirty').

Even pupils who will agree that 3 and 3.0 are the same are often less confident about 0.3 and 0.30.

E **Decimals of a metre** (p 203)

> Sheet 224, metre rules
> Optional: tape measure marked in decimals of a metre

Practical activities

Most will be gained from this section by the use of practical measuring activities. These activities should include measurements where the tenths digit is 0, as in 2.08 m.

Long tape measures, as used by PE departments, often have centimetre scales which are marked in intervals of 0.1 m, for example 2.10, 2.20 etc. These may be useful in illustrating the idea of decimals of a metre.

Some suggested activities are:

Cubits

The ancient measurement of a cubit was said to be the distance between the elbow and the tip of the middle finger. Use blocks of wood or books on a desk at the elbow and fingertips to measure accurately the distance between them to the nearest hundredth of a metre. The class could discuss how consistent a measurement this was.

Standing high jump

First measure the 'reach' of pupils from the floor to their outstretched arms, in metres to two decimal places. Then ask pupils to stand in front of a wall holding a piece of chalk and jump to make a mark as high as they can on the wall. The heights jumped can also be recorded as decimals of a metre. If pupils work in small groups they can list their group's jumps in order of size. The difference between their arms' outstretched 'reach' and the jump height can be used as a fairer assessment of pupils' jumping abilities. A standing long jump event could also be staged.

reach

Records

A tape measure could be laid out on a floor (or pinned to a wall). Results from a recent athletics long jump or historical records could be written on arrow-shaped cards and placed as appropriate on the line. For example some recent winning distances in the Olympic women's long jump were

1948	5.69 m	1976	6.72 m
1952	6.24 m	1980	7.06 m
1956	6.35 m	1984	6.96 m
1960	6.37 m	1988	7.40 m
1964	6.76 m	1992	7.14 m
1968	6.82 m	1996	7.12 m
1972	6.78 m		

Discussion points

◊ The three bulleted questions on the pupil's page raise crucial points for the understanding of decimals.

Height and armspan

◊ This gives further practice in measuring and recording. Measuring needs to be to two decimal places as the two measurements are close.

F Ordering decimals (p 205)

G Place value (p 206)

> Optional: large cards with the digits 0 to 9 on them

The activity below was used in unit 29, 'Calculating with decimals', section F but can extended to two decimal places.

Give each pupil a card with a digit on it. Rule large columns on the board labelled hundreds, tens, units, tenths and hundredths. Give the pupils a number, say 23.46. Invite pupils to come forward and hold the cards in the correct column. Then ask them to make a new number by adding say 0.01 to the number they have already. For example the pupil holding the 6 in the hundredths column would need to be replaced by a pupil holding a 7.

Alternatively this could be done by writing the numbers in columns on the board or OHP and asking pupils to carry out the additions by only changing one digit.

Ⓐ Tenths review (p 198)

A1 (a) 2.3   (b) 1.4   (c) 0.6
     (d) 3.9   (e) 3.4   (f) 4.5

A2 2.5

A3 3

A4 0.9

A5 0.9, 2.8, 3, 4.6, 5

A6 3.3

A7 (a) 3.7 (b) 8.2 (c) 7.6 (d) 7.6

A8 (a) 3.4 (b) 5.6 (c) 5.7 (d) 2.6

A9 (a) 24.6   (b) 33.6   (c) 23.7
     (d) 155.6   (e) 145.7   (f) 9.9

A10 (a) 3 tens or 30   (b) 7 tenths or 0.7
      (c) 4 tenths or 0.4

Ⓑ Tenths and hundredths (p 199)

B1 (a) 0.26 (litre)   (b) 0.53
    (c) 0.63   (d) 0.88
    (e) 0.11   (f) 0.97

B2 (a) 0.13   (b) 0.24   (c) 0.29
    (d) 0.34   (e) 0.38   (f) 0.45

B3 (a) 0.25   (b) 0.15
    (c) 0.35   (d) 0.65

B4 (a) 2.64 cm   (b) 2.18

B5 (a) 6.23   (b) 6.37   (c) 6.42
    (d) 6.47   (e) 6.51   (f) 6.58

B6 (a) 6.25   (b) 6.45
    (c) 6.55   (d) 6.75

B7 (a) 4.51   (b) 4.58   (c) 4.65
    (d) 4.72   (e) 4.8

Ⓒ Less than 0.1 (p 201)

C1 (a) 0.04   (b) 0.07
    (c) 0.02   (d) 0.05

C2 (a) 0.01   (b) 0.05
    (c) 0.09   (d) 0.17
    (e) 0.24   (f) 0.33
    (g) 0.4   (h) 0.45

C3 (a)

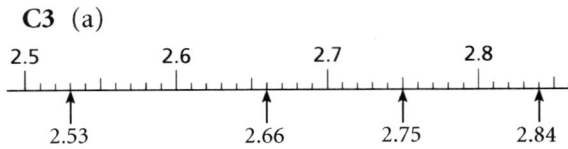

(b)

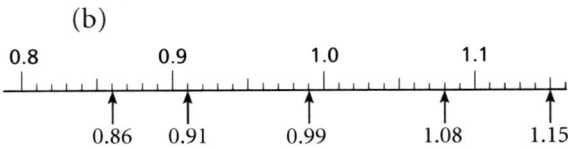

(c)

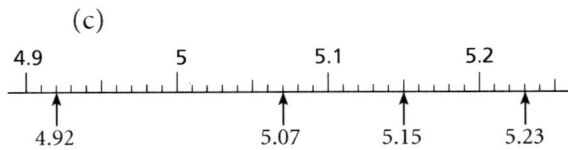

(d)

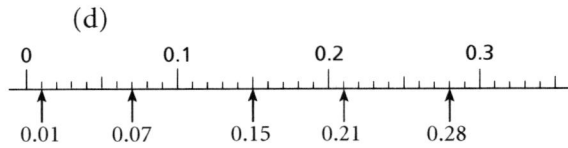

C4 (a) 0.05   (b) 0.09   (c) 0.01

Ⓓ **0.3 and 0.30 are equal** (p 202)

D1 (a) 0.70   (b) 0.10   (c) 0.8
    (d) 0.4

D2 (a) 0.75   (b) 0.95   (c) 0.05

D3 0.05, 0.3 , 0.5

D4 (a) 0.2, 0.65, 0.9
    (b) 0.08, 0.15, 0.4
    (c) 0.07, 0.1, 0.3, 0.62

D5 (a) 0.03, 0.25, 0.6, 0.61
    (b) 0.06, 0.1, 0.17, 0.3
    (c) 0.08, 0.5, 0.55, 0.6

Ⓔ **Decimals of a metre** (p 203)

E1 (a) 3.25 m   (b) 4.68 m

E2 (a) 1 m 39 cm   (b) 1.39 m

E3 (a) 1.7(0) m   (b) 1.07 m
    (c) 2.45 m   (d) 3.06 m

E4 (a) 1.47 m   (b) 1.07 m   (c) 1.6(0) m

E5 The three rows of the table are:

Alan	167 cm	**1 m 67 cm**	**1.67 m**
Kira	**108 cm**	1 m 8 cm	**1.08 m**
Greg	**140 cm**	**1 m 40 cm**	1.4 m

E6 (a) Cobra (3.05 m)
    (b) Boa constrictor (3.3 m)

E7 1.18 m, 1.4 m, 1.66 m

E8 0.09 m, 0.53 m, 0.7 m

E9 (a) 4.62 m   (b) 4.65 m
    (c) 4.68 m   (d) 4.73 m

E10 (a) 7.01 m   (b) 7.03 m
    (c) 7.07 m   (d) 7.11 m

E11 (a)

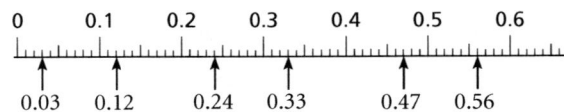

(b)

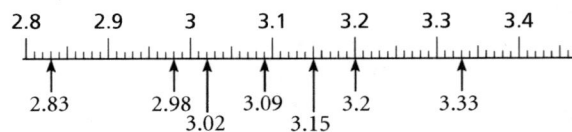

(c)

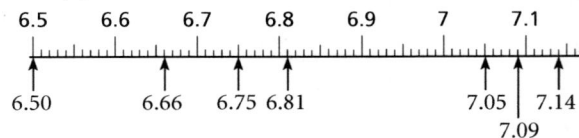

(d)

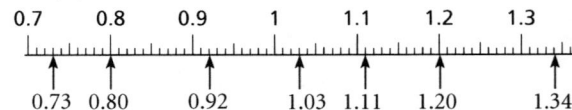

Ⓕ **Ordering decimals** (p 205)

F1 2.35, 2.4

F2 5.17, 5.03, 5.1

F3 (a) 5.09, 5.32, 5.76, 5.84, 6
    (b) 2.8, 3.07, 3.19, 3.2, 3.5
    (c) 0.15, 0.45, 0.6, 1.07, 1.1

F4 HOLIDAYS

F5 ENJOYABLE

F6 DINOSAUR

F7 (a) 4.2 m   (b) Barker
    (c) Church   (d) 4.5 m

Ⓖ **Place value** (p 206)

G1 (a) 7 hundreds or 700
    (b) 2 thousands or 2000
    (c) 3 units or 3
    (d) 1 ten or 10
    (e) 8 hundredths or 0.08

G2 (a) 2723.58   (b) 3713.58
    (c) 2713.68   (d) 2714.58
    (e) 2713.59   (f) 2813.58

G3 (a) 1526.35   (b) 1427.35
    (c) 1426.45   (d) 2426.35
    (e) 1436.35   (f) 1426.36

G4 (a) 0.7   (b) 0.03   (c) 60
    (d) 0.8   (e) 0.04   (f) 0.5

G5 60

G6 0.7

G7 0.03

What progress have you made? (p 207)

1 (a) 4.84 (b) 4.91 (c) 5.02 (d) 5.08
    (e) 0.06 (f) 0.14 (g) 0.19 (h) 0.28

2 (a) 1.45 m   (b) 2.03 m
    (c) 3 m 20 cm

3 (a) 6.68, 6.7, 6.9, 7.04, 7.3
    (b) 0.07, 0.1, 1.56, 3.24, 4

4 (a) 8 tenths or 0.8
    (b) 4 hundredths or 0.04
    (c) 7 hundredths or 0.07

Practice booklet

Sections B and C (p 83)

1 (a) 0.34 (litre)   (b) 1.26
    (c) 0.98   (d) 0.03
    (e) 3.51

2 (a) 0.14   (b) 0.23   (c) 0.28
    (d) 0.32   (e) 0.37   (f) 0.43

3 (a) 3.94   (b) 3.99   (c) 4.02
    (d) 4.07   (e) 4.15   (f) 4.22
    (g) 4.26   (h) 4.31

4 (a) 4.15   (b) 4.25   (c) 4.05
    (d) 3.95

5

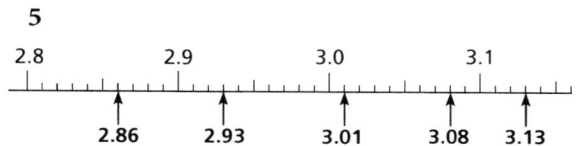

Section D (p 84)

1 (a) 0.55   (b) 0.25
    (c) 0.05   (d) 0.45

2 A and H (0.7 and 0.70)
    B and G (0.5 and 5 tenths)
    C and F (5 hundredths and 0.05)
    D and E (0.07 and 7 hundredths)

3 (a) 0.04, 0.11, 0.18, 0.20, 0.5
    (b) 0.30, 0.34, 0.4, 0.59, 0.80
    (c) 0.07, 0.21, 0.40, 0.54, 0.9
    (d) 0.05, 0.1, 0.17, 0.3, 0.6

4 (a) 0.26, 0.31, 0.40, 0.5, 0.53
    (b) 0.20, 0.26, 0.59, 0.6, 0.80
    (c) 0.09, 0.2, 0.39, 0.9, 0.99
    (d) 0.25, 0.3, 0.81, 0.87, 0.9

5 (a) 2.07, 2.39, 2.5, 2.8
    (b) 1.01, 1.1, 1.43, 1.5

Section E (p 85)

1 (a) 3.27 m (b) 1.56 m
   (d) 5.64 m (e) 4.98 m
   (f) 2.09 m

2 A and H (1 metre and 8 cm, 1.08 m)
   B and D (1.3 m, 130 cm)
   C and G (1.8 m, 1 metre and 80 cm)
   E and F (1 metre and 3 cm, 1.03 m)

3 (a) 2.09 m
   (b) 3.20 m or 3.2 m
   (c) 4.80 m or 4.8 m
   (d) 9.01 m
   (e) 6.40 m or 6.4 m
   (f) 6.04 m

4

Name	Height in cm	Height in metres and cm	Height in metres
Mel	152 cm	**1 m 52 cm**	**1.52 m**
Ginger	**149 cm**	1 m 49 cm	1.49 m
Ali	**159 cm**	**1 m 59 cm**	1.59 m
Kay	**109 cm**	1 m 9 cm	**1.09 m**
Morag	**160 cm**	**1 m 60 cm**	1.6 m

5 (a) 2.87 m, 2.9 m, 3.08 m, 3.1 m
   (b) 0.05 m, 0.5 m, 0.9 m, 1.01 m

6 B, C

Section F (p 86)

1 6.3, 6.17 and 6.37

2 (a) 4.09, 4.24, 4.25, 4.3, 5
   (b) 0.08, 0.65, 0.7, 1, 1.5

3 TRIANGLE

4 6.06 or 6.48

5 (a) Jamie White
   (b) Pete Smith
   (c) Joe Hall (1.5 m)
        Suneet Patel (1.48 m)
        Garry Fulton (1.59 m)
        Jamie White (1.6 m)

Section G (p 87)

1 (a) 700 or 7 hundreds
   (b) 0.3 or 3 tenths
   (c) 1
   (d) 0.09 or 9 hundredths

2 A, S; B, R; C, T; D, U;
   E, Q; F, P; G, R

3 (a) 63.7 (b) 72.7 (c) 62.9
   (d) 38.74 (e) 33.74 (f) 39.64
   (g) 39.73 (h) 30.1

4 (a) 458.1 + **10** = 468.1
   (b) 34.91 + **0.01** = 34.92
   (c) 103.6 + **50** = 153.6
   (d) 75.32 − **1** = 74.32
   (e) 54.8 − **0.8** = 54
   (f) 587.5 + **400** = 987.5
   (g) 6.77 − **0.7** = 6.07
   (h) 6.77 − **0.07** = 6.7

㉞ Practical problems

These problems use equipment so need to be done in class. Pupils need not do all the tasks: just doing some will tell you a lot about how well they can measure, estimate and apply number skills in problem solving.

One approach is to set up a 'circus' of tables; on each table is the equipment for one task and a label giving its name and page number (tasks with readily available equipment can be duplicated on more than one table). Pupils move round the circus following the instructions in the pupil's book. It is a good idea to have some 'exercise' work ready in case there is a log-jam as pupils go around the circus.

Alternatively, you can set up just one task (possibly in duplicate) and, while the rest of the class get on with written work, individuals or small groups come out in turn to do it.

'Weighty problems' is for a small group of pupils; the others can be done individually or in pairs.

Many pupils find weighing difficult, whether interpreting scale graduations on mechanical scales or coping with decimal places on digital balances. So it may be a good idea to add some straightforward weighing to the collection of tasks.

Weighty problems (p 208)

> Scales or electronic balance
> Two stones, say about 4 cm and 8 cm in diameter
> A collection of familiar objects, including one with its weight clearly marked on it (for example a 500 g or 1 kg bag of sugar)

◊ In task 1 each group's estimates could be displayed on a dot plot (if pupils are familiar with them) and ideas of spread and over- or under-estimation discussed by the class. It is a good idea if everybody in the group checks the reading on the scales when the stones are weighed.

◊ After task 2 pupils could discuss ways of deciding which estimated order of weights was the best.

A related activity that goes well is for a pupil to hold the object of known weight (bag of sugar or whatever) in one hand and a different object in the other; the pupil estimates the weight of the other object then checks by weighing.

Beans (p 208)

> Two identical large sweet jars with lids, one empty and the other at least half filled with dried beans or pasta shapes with its lid taped down (butter beans are suitable, but not red kidney beans or other varieties that are poisonous when uncooked)
> About 100 extra beans or pasta shapes of the type in the jar
> An electronic balance or scales sensitive enough to weigh a few grams

◊ These are two approaches pupils have used to start solving the problem.

- Putting a layer of the extra beans into the empty jar and measuring the layer's height.
- Finding the weight of the beans in the jar by subtracting the weight of the empty jar.

In the second case, some go on to weigh a single bean. If so, ask them to check whether the beans all weigh the same. If they don't weigh the same, can pupils suggest a way to deal with this?

Cornflakes (p 209)

> A full box of cornflakes, with its price
> A cereal bowl
> An electronic balance or scales

◊ Some pupils may need a hint to work out the weight of the cornflakes without the bowl.

◊ You could extend the work to comparing the cost of different types of cereal or comparing the cost of an individual portion box with that of the same amount of cereal in a full-size box.

Getting better (p 209)

> A 5 ml spoon labelled 5 ml
> A container with a scale graduated in ml
> Three different sized medicine bottles (one less than 60 ml) distinguished by colour or labelling, but without their capacities marked
> Water, a tray and some paper towels

◊ A prepared answer sheet may help weaker pupils.

◊ Follow-up might include

- discussion of the appropriate level of accuracy
- estimation by pupils, perhaps as a homework assignment, of their daily fluid intake

Children's TV (p 210)

Radio Times, or other paper with TV programme listings

◊ Some pupils may need to be reminded that there are 60 minutes in an hour.

◊ Pupils should be encouraged not to take easy options such as listing six half-hour programmes.

◊ Some teachers have asked their pupils to write a letter to their friend abroad to explain why they chose the programmes.

Follow-up for some pupils might include writing out a schedule for their tape, starting at 00:00.

Windfall (p 210)

A shopping catalogue such as Argos or Littlewoods Optional: an order form for the pupils to fill in adds to the realism of the task

Be aware of pupils whose interest lies mainly in the contents of the catalogue, rather than the task in hand.

③⑤ Amazing but true! (p 211)

This unit revises work on coordinates.

The maze shown in the picture, based on a castle, was designed and
made by Adrian Fisher, a professional maze designer.

Essential

Squared paper

◊   You might wish to use the maze shown on page 211 with the whole class
before pupils design their own mazes.

◊   An early check that all pupils are giving *x*-coordinates first is essential.
Instead of using compass directions to give instructions pupils could use
'left' and 'right' to practise the visualisation skills used in Finding your
way (section D of unit 1 'First bites').

Start at (0, 1)	Go east
At (9, 1)	Go north
At (9, 7)	Go west
At (8, 7)	Go south
At (8, 2)	Go west
At (2, 2)	Go north
At (2, 5)	Go east
At (3, 5)	Go south
At (3, 3)	Go east
At (7, 3)	Go north
At (7, 5)	Go west
At (6, 5)	Go north
At (6, 6)	Go west
At (5, 6)	Go north
At (5, 7)	Go west
At (3, 7)	Go north
At (3, 9)	Go east, to the exit

36 Is it an add?

In this unit pupils think about, discuss and choose appropriate operations and calculations for number problems. Although they may be able to solve contextual problems where the operation is clear, pupils often have difficulty choosing the right operation and calculation for a number problem.

p 212 **A** Add, subtract, multiply or divide?	Discussing the choice of the appropriate operation for a problem
p 216 **B** Video cassettes	Choosing an appropriate calculation
p 217 **C** Telling tales	Choosing calculations for written questions
p 217 **D** Check it out	Checking answers from a calculator

Essential

Calculators

Practice booklet pages 88 to 91

A Add, subtract, multiply or divide? (p 212)

Pupils discuss and select the correct operations for given problems.

◊ Pages 212 and 213 could be used for an introductory class discussion. First ask pupils, working in pairs or small groups, to decide, for each picture question on page 212, whether they would add, subtract, multiply or divide. They could try to explain why the operations they choose are correct.

You should try to bring out the features of each picture – in particular for multiplication

'We used these ideas and worked through these pages together, which they all enjoyed, and they liked giving answers.'

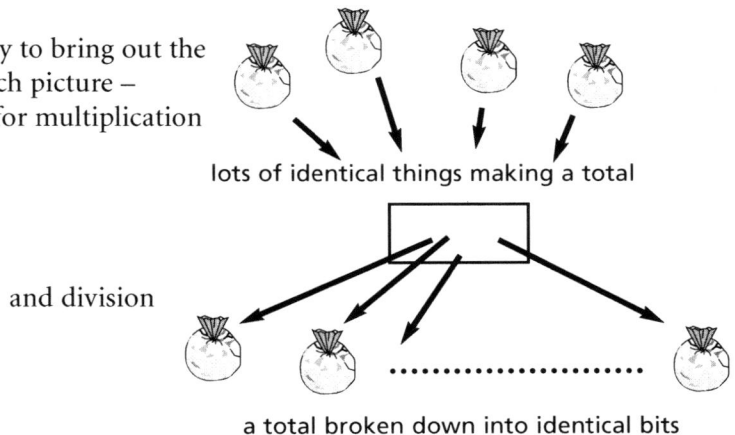

lots of identical things making a total

and division

a total broken down into identical bits

Page 213 leads into the rest of the unit by asking which calculation from a set they would do to find the answer to each question. Ask them to try to explain why the calculation they have chosen for each question is correct.

◊ Encourage discussion about pages 214 and 215. In some cases there may be more than one correct answer, for example 20 × 3 and 3 × 20.

B **Video cassettes** (p 216)

◊ Make it clear to the pupils that they have to write the calculations, not the answers (for example, '5 × 3' for B1, not '15').

C **Telling tales** (p 217)

C3 Pupils may find writing their own tales difficult. They could write these as questions like C1 and ask other pupils to do them.

D **Check it out** (p 217)

Calculators

The objective of this section is to encourage pupils to use simple checking procedures when they have used a calculator for a calculation.

◊ In the initial discussion pupils may come up with their own methods for checking. The most easily applied checks are:
  • Reversing the order – with additions, particularly long lists, and multiplication. You will need to point out that this cannot be used with subtraction and division.
  • Inverse operations – for example checking a division by multiplying the answer by the divisor. Inverse operations can be used to check all one-step calculations, and their use will help to develop an understanding of inverses.

Challenge (p 219)

You may wish to point out that 1000 kg is a metric tonne.

The optimum solution is parcels B, C and F whose combined weight is 371 + 535 + 89 = 995 kg. (With appropriate pupil's check of course!)

Ⓐ Add, subtract, multiply or divide? (p 212)

A1 8 × 20 or 20 × 8

A2 80 ÷ 8

A3 80 ÷ 20

A4 20 × 12

A5 12 × 8 or 8 × 12

A6 12 + 8 or 8 + 12

A7 12 × 8 or 8 × 12

A8 120 ÷ 12

A9 12 − 8

A10 20 − 8

A11 30 × 12 or 12 × 30

A12 30 − 12

A13 120 ÷ 30

Ⓑ Video cassettes (p 216)

B1 5 × 3 or 3 × 5

B2 6 ÷ 4

B3 6 + 16

B4 16 − 14

B5 4 × 14

B6 11 − 8

B7 5 + 8

B8 5 × 8

B9 4 − 1.5

B10 15 ÷ 3

Ⓒ Telling tales (p 217)

C1 20 ÷ 4

C2 (a) 35 − 23   (b) 96 ÷ 12
      (c) 15 + 8   (d) 18 × 5

C3 The pupil's tales

Ⓓ Check it out (p 217)

In all the questions in this section, pupils should demonstrate that they have carried out an appropriate check.

D1 (a) 731   (b) 783   (c) 3404
      (d) 167   (e) 132   (f) 2368
      (g) 11 376   (h) 22   (i) 6

D2 (a) 267 + 86 + 3825 + 228; 4406 kg
      (b) 267 − 228; 39 kg
      (c) 267 × 8; 2136 kg

D3 237 − 128; 109 cm

D4 414 ÷ 18; 23 pipes

D5 (a) 137 + 141 + 131 + 137; 546 litres
      (b) Daisy
          Totals are:
          Abigail   546 litres
          Buttercup   522 litres
          Cowslip   545 litres
          Daisy   563 litres
      (c) 2nd week
          Totals are:
          1st week   537 litres
          2nd week   551 litres
          3rd week   543 litres
          4th week   545 litres

D6 (a) 27 × 36; 972 windows
      (b) 396 ÷ 22; 18 floors

D7 (a) 5.62 × 38; £213.56
      (b) 241.66 ÷ 5.62; 43 hours

What progress have you made? (p 219)

1 (a) 6 × 12   (b) 150 ÷ 6
    (c) 30 − 12   (d) 150 + 30

2 6 × 4

3 (a) 735 with pupil's check
    (b) 1015 with pupil's check

Practice booklet

Section B (p 88)

Soft drinks

 1 $12 - 5$ **2** $6 + 12$

 3 3×3 **4** $2 + 3$

 5 12×4 **6** $5 - 3$

 7 $30 \div 6$ **8** $12 \div 2$

Films

 1 $2 + 3$ **2** $6 - 2$

 3 $6 - 2$ **4** 3×8

 5 36×2 **6** $36 - 24$

 7 3×3 **8** $180 \div 36$

 9 $8 \div 2$ **10** $18 \div 6$

Section C (p 90)

 1 (a) 12×6 (b) $120 \div 6$

  (c) $15 + 6$ (d) $24 - 5$

 2 (a) 40×7 (b) $24 - 8$

  (c) $420 \div 60$ (d) $15 + 17$

  (e) $2000 \div 400$ (f) 64×25

 3 The pupil's tales

Section D (p 91)

For all questions in this section, pupils should show the calculation they did to check.

 1 (a) 931   (b) 810   (c) 194

  (d) 1330   (e) 1222   (f) 12 492

  (g) 27   (h) 51

 2 (a) $578 + 1253 + 1820 + 2176 + 987$;
   6814 vouchers

  (b) $2176 - 1253$; 923 vouchers

  (c) $1820 \div 28$; 65 each

 3 (a) 144×15; 2160 sweets

  (b) $144 - 87$; 57 tubes

  (c) 40×55; 2200 sweets

  (d) A box of bags

③⑦ Graphs and charts

Pupils interpret graphs and charts from real-life sources and draw their own frequency bar charts and line graphs.

Essential	**Optional**
Sheets 182 to 185	Wall's Monitor
	OHP transparency ruled as pupils' graph paper
	Sheet 186
Practice booklet pages 92 to 95	

🅐 **Children's income** (p 220)

> Optional: Wall's Monitor (Available from Birds Eye Wall's Limited, Station Avenue, Walton-on-Thames, Surrey KT12 1NT; tel. 01932 263000. This was free at the time of writing.)

'Very good, relevant and they all have an opinion.'

'Wall's info led to considerable useful discussion and caught the pupils' interest.'

The chart shows the total weekly income of children, which is the sum of pocket money, hand-outs from other relatives and any earnings. Before you begin you may want to collect data from the class to compare with the data given in the chart. This can be a sensitive issue so try to avoid a show of hands: getting pupils to write their income on a piece of paper that you collect can alleviate problems.

The data from Wall's Monitor could be used when drawing graphs and charts later in the unit. Wall's Monitor is an annual pocket money survey and is based on data provided by Gallup. It provides trend data on children's income. It contains a host of data.

B **Shut up!** (p 221)

Comparing the information on the two graphs in this section can lead to a useful class discussion.

C **Off the record** (p 222)

T

'*This worked well as a group activity.*'

You may need to explain what an LP or a single is.

D **Equal shares** (p 223)

E **Drawing graphs and charts** (p 224)

Sheets 182 to 185
Optional: OHP transparency as pupils' graph paper, sheet 186

T

◊ You can use the data on the top of page 224 to introduce drawing a frequency bar chart. You may wish to discuss suitable intervals for grouping, but pupils are not expected to choose their own groupings in this unit (intervals of 0–9, 10–19 etc. will be fine). Note that this may be the first time that pupils have met the idea of grouping data.

If you use an OHP, you will find it useful to have a transparency ruled just like the graph paper the pupils use. Otherwise, a grid on the board is essential.

You will need to point out how to draw suitable bars on the chart.
Either label each bar '0–9', '10–19', leaving gaps between bars to emphasise the discreteness, or mark off a normal continuous scale.
If a data item falls on a boundary between class intervals, a convenient rule is to include it in the upper interval.

No distinction has been made in this unit between representing discrete and continuous data.

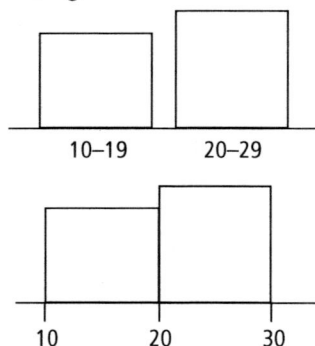

◊ Only a few data sets are given in the pupils' book, as the work is more motivating if you use data relevant to the class.
You could use their own scores from a recent test, heights, reaction times or any data which can be grouped into convenient intervals.

◊ Sheets 182 and 184 provide follow-up examples which can be used individually or worked through with the teacher.
Sheets 183 and 185 are blanks for your own data.

◊ This is a tally chart for the data at the top of page 224.

Score	Tally	Frequency
0–9	/	1
10–19	///	3
20–29	////	4
30–39	ЖЖ ЖЖ /	11
40–49	ЖЖ	5
50–59	ЖЖ /	6
60–69	////	4
70–79	ЖЖ	5
80–89	/	1
	Total	40

E1–3 Sheet 186 may be useful for questions E1 to E3 if pupils find tallying difficult. One pupil can read out each data item and cross it out on the sheet, while another pupil makes a corresponding tally mark in the table.

E4, 5 These questions are best done on squared paper. If pupils find drawing the scales difficult, you might draw them yourself on squared paper and copy your sheet for pupils to use.

𝔸 **Children's income** (p 220)

A1 (a) 605p   (b) Scotland
(c) North West
(d) Difference = 605p – 397p = 208p

A2 (a) 50%   (b) 25% to 30%

A3 (a) Sweets   (b) Sweets
(c) Cigarettes

A4 4629 pupils (written at bottom)

A5 The pupil's answers

𝔹 **Shut up!** (p 221)

B1 (a) 1990
(b) (i) 1987 and 1988 (ii) About 60 000
(c) 1985 and 1986 (d) 1989 and 1990

B2 12 000 (roughly)

B3 For example, 'Between 1983 and 1985 complaints rose steadily. Then they dropped between 1985 and 1986. They rose in the next year and stayed the same between 1987 and 1988. They rose again between 1988 and 1989 and then rose very steeply up to 1990.'

B4 The scale is not continuous from zero.

B5 Between 1985 and 1986

B6 1986

ℂ **Off the record** (p 222)

C1 (a) The sales of CDs started in 1983 and went up every year.
(b) Sales of cassettes rose to just over 80 million in 1989, then dropped.
(c) 1989   (d) 1986
(e) About 80 million

C2 The pupil's questions.

Ⓓ Equal shares (p 223)

D1 (a) Repairs (b) Washing and ironing

D2 Washing and ironing – it has the lowest percentage of 'mainly men' and 'shared equally' put together.

D3 Washing and ironing, cleaning house and making evening meal

D4 Disciplining children

D5 Washing dishes, paying bills and perhaps disciplining children all have roughly equal percentages of 'mainly men' and 'mainly women'.

Ⓔ Drawing graphs and charts (p 224)

E1

Age group	Tally	Frequency
0–9	/	1
10–19	//	2
20–29	⦀⦀ /	6
30–39	////	4
40–49	////	4
50–59	⦀⦀	5
60–69	⦀⦀	5
70–79	////	4
80–89	/	1
	Total	32

The pupil's bar chart

E2

No. of cars	Tally	Frequency
50–59	⦀⦀ /	6
60–69	⦀⦀	5
70–79	////	4
80–89	⦀⦀ ⦀⦀	10
90–99	⦀⦀ ///	8
100–109	////	4
110–119	///	3
	Total	40

The pupil's bar chart

E3

Reaction time	Tally	Frequency
10–12	⦀⦀	5
13–15	⦀⦀ //	7
16–18	⦀⦀ ⦀⦀ //	12
19–21	////	4
22–24	/	1
25–27	/	1
	Total	30

The pupil's bar chart

E4, 5 The pupil's graphs

178 • 37 Graphs and charts

Sheet 182

Age group	Tally	Frequency
0–9	///	3
10–19	⦀⦀ /	6
20–29	⦀⦀ //	7
30–39	⦀⦀ /	6
40–49	⦀⦀ /	6
50–59	///	3
60–69	⦀⦀ /	6
70–79	//	2
80–89	/	1
	Total	40

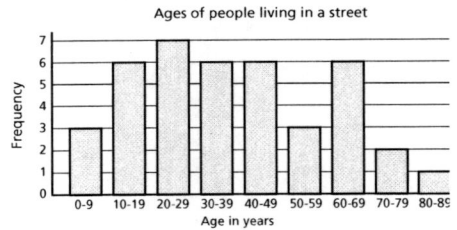

Ages of people living in a street

Sheet 184

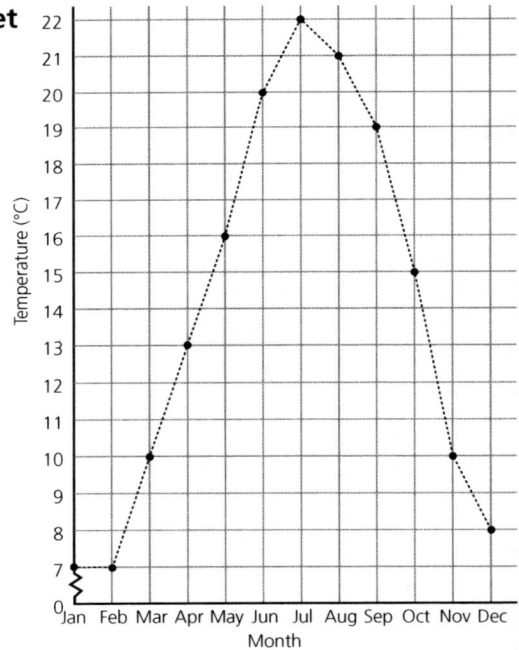

What progress have you made? (p 225)

1 (a) 13 (b) 8

2 (a) 38°C

(b) It was 37°C at 3 p.m., then rose to 40°C at 6 p.m., dropping gradually to 38°C at 9 p.m. with a slight rise in between.

3

Babies born in hospital

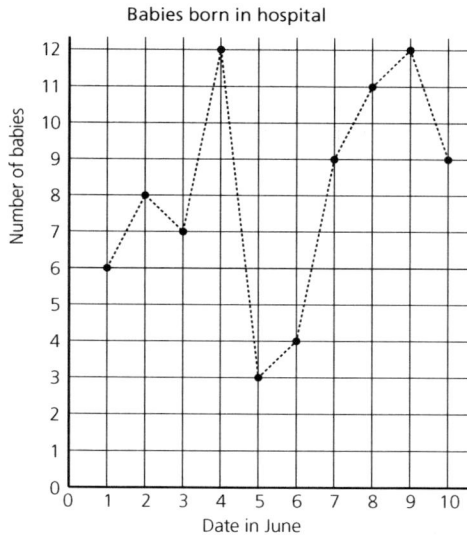

4

Age group	Tally	Frequency
0–9	///	3
10–19	////	4
20–29	―//// /	6
30–39	///	3
40–49	//	2
50–59	/	1
60–69	////	4
70–79	////	4
	Total	27

5

Ages of people living in a village

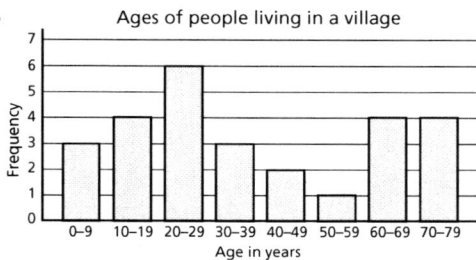

Practice booklet

Section A (p 92)

1 (a) Bus   (b) Car   (c) 30%

    (d) 8%   (e) Car   (f) Bike

    (g) More than year 7

    (h) More than year 7

Section C (p 93)

1 (a) London, 4°C   (b) Sydney, 22°C

    (c) The temperature went down until the 7th and 8th; then went up until the 11th, and then went down again.

    (d) The temperature went up until the 5th and 6th; then went down until the 10th and 11th; then went up again.

    (e) 10°C   (f) 14°C

    (g) On the 4th, 5th, 6th, 7th, 8th, 13th, 14th and 15th

Section D (p 94)

1 (a) The UK   (b) Greece

    (c) Portugal and Greece

    (d) The Netherlands

    (e) Germany

    (f) Spain, Portugal, Italy and Germany

    (g) Spain, Portugal, Italy, Ireland, Greece and Germany

Section E (p 95)

1 (a)

Weight (g)	Tally	Frequency
0–9	/	1
10–19	―//// /	6
20–29	―////	5
30–39	―//// ///	8
40–49	―//// /	6
50–59	////	4
	Total	30

    (b) The pupil's bar chart

    (c) The most common weight of fish caught in this pond is between **30** and **39** grams.

2 (a) The pupil's graph

    (b) The rainfall is very low from December to April, then increases to June, drops a little in July, and goes up in August to October; then it drops once more.

The information on the pupil's page is intended for use as a context for orally given questions.

General advice on oral work is given in the notes for unit 8, 'Oral questions: calendar'.

'It was the last 15 minutes of the lesson and lunch time was approaching. Most pupils did very well making only one or two small mistakes or getting all correct. It was very good diagnostically.'

These sample questions are roughly in order of increasing difficulty.

1	Find four items which cost under £1 each.	Note pad, pencils, ruler, eraser
2	What is the total cost of a note pad and an eraser?	£0.72 or 72p
3	How much change from a £10 note do I get if I buy the stapler?	£2.20
4	Find the cost of four packs of ink cartridges.	£4.28
5	A Stick-it note measures 4 cm by 5 cm. Draw it accurately.	
6	Find the cost of a pencil case and an eraser.	£1.25
7	Pencil cases go up in price by 10p. How much does one cost now?	£1.20
8	What is the most expensive item?	Hole punch
9	Which item costs £0.79?	Ruler
10	Find the cost of four note pads.	£2.28
11	Find the cost of ten rulers.	£7.90
12	The shop has a half-price sale. What is the sale price of the hole punch?	£4.25

㊴ Working with fractions

This unit revises finding fractions of numbers, starting with a problem-solving activity for the whole class.

p 228 **A** Chocolate	Problem-solving activity
p 228 **B** Simple fractions of numbers	Working out fractions such as $\frac{1}{2}$ of 1972
p 230 **C** Other fractions of numbers	Working out fractions such as $\frac{2}{3}$ of 792
p 231 **D** Improper fractions and mixed numbers	

Essential	**Optional**
6 bars or blocks of something, which can be divided up and shared out equally	Bars of chocolate
Practice booklet pages 96 to 99	

𝔸 Chocolate (p 228)

'I thought this would be chaos and pupils would gain little from it. In fact, lots of useful discussion was generated! Had I remembered to buy the chocolate it would have been even better. (We used sheets of paper to represent chocolate.)'

This is a problem-solving activity in which pupils can apply their understanding of fractions. Pupils decide which table to sit at in the hope of getting the most chocolate.

The activity provides opportunities for

- introducing pupils to the idea that a fraction like $\frac{2}{3}$ of a bar can mean '2 bars divided between 3 pupils'
- introducing methods of finding fractions of numbers such as $\frac{2}{3}$ of 24

6 bars or blocks of something, which can be divided up and shared out equally
Optional: bars of chocolate

Getting started

◊ It is best to start with a fairly simple situation. For example:

- Distribute 3 chocolate bars on 2 tables as shown.
- Choose a group of, say, 8 pupils to take part.
- Tell them that you will ask them one by one to choose a table to sit at (they cannot change their minds later).
- Once they are all seated, they will get an equal share of the chocolate on their table.

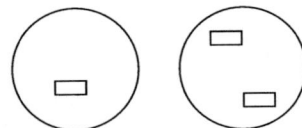

Different groups of pupils can be chosen to give all pupils a chance to take part.

Solving the problem

◊ As pupils choose where to sit, involve the whole class and ask questions such as:

- Where do you think the next person should sit? How would you advise them?
- What fraction of a chocolate bar would each person get at this table if no one else sits here? (Emphasise that 2 bars divided between 3 people means that each person gets $\frac{2}{3}$ of a bar.)

◊ Once the last pupil has chosen, ask pupils to decide which pupils get the most chocolate and to justify their decision. Pupils could consider this in small groups and then explain how they decided to the whole class.

Variations

- Change the number of pupils who are to sit at the tables.
- Change the number of tables and/or the numbers of bars on each table. For example, use three tables – one with 1 bar, another with 2 and the third with 3.
- Once a group is seated at the tables, and who gets most is decided, ask if a different seating arrangement might make the division of the bars fairer.

Follow-up

The activity works well when pupils cannot see how many squares are in each chocolate bar. However, repeating the activity when pupils **can** see how many squares are in a bar will provide a useful lead-in to the rest of the unit. Choose (or draw) bars of chocolate that have a number of squares with plenty of factors, such as 24.

Ⓑ Simple fractions of numbers (p 228)

You could give pupils some oral fractions questions, such as $\frac{1}{2}$ of 42, $\frac{1}{4}$ of 100, $\frac{1}{6}$ of 24, $\frac{1}{8}$ of 48 etc. to work out as quickly as possible.

C Other fractions of numbers (p 230)

T
Some teachers have found it helpful if pupils lay out their calculations as in the two examples at the top of page 230.

D Improper fractions and mixed numbers (p 231)

T
Pupils who find this difficult may be aided by using quarter circles, as in the diagram in the introduction. Circles cut into thirds, sixths etc. may be used to extend the work.

It will remind pupils of the vocabulary of fractions if, from time to time, you ask them oral questions such as
- 'What is nine quarters as a mixed number?'
- 'Write two and a half as an improper fraction.'

B Simple fractions of numbers (p 228)

B1 (a) 2   (b) 7   (c) 11
(d) 5   (e) 18

B2 (a) 2   (b) 8   (c) 6
(d) 10   (e) 13

B3 P and D   Q and C   R and E
S and B   T and A

B4 The pupil's diagrams

B5 (a) 8   (b) 7   (c) 9
(d) 2   (e) 9

B6 (a) 782   (b) 493   (c) 7869
(d) 52 341   (e) 259   (f) 25 897
(g) 129 670   (h) 456   (i) 633

B7 About 3496

B8 500 000

B9 8000

B10 12 500

C Other fractions of numbers (p 230)

C1 (a) 18   (b) 24   (c) 30
(d) 3   (e) 33

C2 (a) 16   (b) 20   (c) 12
(d) 14   (e) 22

C3 P and D   Q and F   R and A
S and C   T and B   U and E

C4 (a) 27   (b) 12   (c) 15
(d) 35   (e) 40

C5 16

C6 (a) 160 cm   (b) 24 years old

C7 (a) 528   (b) 3688
(c) 2800   (d) 369 000

C8 128 m

C9 300 000

C10 (a) 68 000 km^2 (b) 340 000 km^2

C11 600 leaves

D Improper fractions and mixed numbers (p 231)

D1 (a) (i) 5   (ii) $1\frac{1}{4}$
(b) $\frac{5}{4} = 1\frac{1}{4}$

D2 (a) 9   (b) $2\frac{1}{4} = \frac{9}{4}$

D3 (a) $3\frac{1}{4} = \frac{13}{4}$ (b) $3\frac{3}{4} = \frac{15}{4}$

     (c) $\frac{17}{4} = 4\frac{1}{4}$ (d) $\frac{21}{4} = \mathbf{5\frac{1}{4}}$

D4 (a) $\frac{5}{3}$ litres   (b) $1\frac{2}{3}$ litres

D5 (a) $\frac{4}{3} = 1\frac{1}{3}$ (b) $2\frac{2}{3} = \frac{8}{3}$

     (c) $2\frac{1}{3} = \frac{7}{3}$ (d) $\frac{10}{3} = \mathbf{3\frac{1}{3}}$

D6 (a) $1\frac{1}{2}$ (b) $2\frac{1}{2}$ (c) $1\frac{3}{5}$

     (d) $2\frac{2}{5}$ (e) $1\frac{3}{7}$

D7 (a) $\frac{7}{2}$ (b) $\frac{13}{2}$ (c) $\frac{25}{3}$ (d) $\frac{20}{3}$ (e) $\frac{15}{7}$

What progress have you made? (p 232)

1 (a) 9   (b) 5   (c) 6

2 (a) 9   (b) 5   (c) 10

3 (a) 55   (b) 200   (c) 126

4 (a) 14   (b) 60   (c) 21

5 (a) 128 688   (b) 2688

     (c) 513   (d) 840

6 (a) $\frac{11}{3} = \mathbf{3\frac{2}{3}}$ (b) $\frac{9}{2} = \mathbf{4\frac{1}{2}}$ (c) $\frac{9}{5} = \mathbf{1\frac{4}{5}}$

     (d) $1\frac{1}{5} = \frac{6}{5}$ (e) $4\frac{1}{3} = \frac{13}{3}$ (f) $1\frac{5}{6} = \frac{11}{6}$

Practice booklet

Section A (p 96)

1 They each get $\frac{1}{4}$ of a bar.

2 $\frac{2}{6}$ or $\frac{1}{3}$ of a bar each

3 $\frac{1}{2}$ of a bar

4 (a) A person at table Q

     (b) They get $\frac{3}{4}$ of a bar each.

5 (a) The people at table Q – they get $\frac{3}{5}$ of a bar each.
         At table P they get $\frac{1}{3}$ of a bar each.

     (b) The people at table Q – they get a whole bar each.
         At table P they get $\frac{1}{2}$ a bar each.

     (c) The person at table Q – he or she gets 2 bars.
         At table P they get $1\frac{1}{2}$ bars each.

Section B (p 97)

1 P and B; Q and A; R and D; S and C

2 A $\frac{1}{3}$ of 12 = 4   B $\frac{1}{5}$ of 15 = 3

   C $\frac{1}{7}$ of 14 = 2   D $\frac{1}{3}$ of 18 = 6

   E $\frac{1}{2}$ of 12 = 6

3 Drawings showing these amounts:

     (a) 5   (b) 5   (c) 8   (d) 4

4 (a) 7   (b) 5   (c) 4   (d) 8

     (e) 7   (f) 9   (g) 7   (h) 6

     (i) 2   (j) 7   (k) 4   (l) 4

Section C (p 98)

1 P and A; Q and E; R and C; S and B; T and F; U and D

2 (a) 27   (b) 36   (c) 18   (d) 16

     (e) 9   (f) 25   (g) 21   (h) 16

     (i) 3   (j) 30   (k) 180   (l) 300

3 (a) 4   (b) 16   (c) 24   (d) 30

4 (a) 10 m   (b) 30 m   (c) 24 m   (d) 25 m

Section D (p 99)

1 (a) 4   (b) $1\frac{1}{3}$

2 $2\frac{1}{3} = \frac{7}{3}$

3 (a) $1\frac{2}{3} = \frac{5}{3}$ (b) $3\frac{1}{3} = \frac{10}{3}$

     (c) $\frac{13}{3} = \mathbf{4\frac{1}{3}}$ (d) $\frac{8}{3} = \mathbf{2\frac{2}{3}}$

4 (a) $1\frac{3}{5} = \frac{8}{5}$ (b) $2\frac{2}{5} = \frac{12}{5}$

     (c) $1\frac{5}{6} = \frac{11}{6}$ (d) $2\frac{1}{6} = \frac{13}{6}$

5 (a) $1\frac{4}{5} = \frac{9}{5}$ (b) $\frac{13}{5} = \mathbf{2\frac{3}{5}}$

     (c) $3\frac{1}{5} = \frac{16}{5}$ (d) $3\frac{3}{5} = \frac{18}{5}$

6 (a) $2\frac{3}{4}$ (b) $3\frac{2}{3}$ (c) $2\frac{1}{5}$ (d) $1\frac{4}{7}$

7 (a) $\frac{15}{7}$ (b) $\frac{27}{8}$ (c) $\frac{35}{6}$ (d) $\frac{47}{7}$

④⓪ 3-D shapes

This unit focuses on identifying the nets of open cubes and visualising 3-D shapes from 2-D drawings. Many other spatial skills are practised in the activities.

T	p 233 **A** Pentominoes	Investigating different patterns with five tiles
	p 234 **B** Nets	Looking at nets for an open cube
T	p 235 **C** Seeing in 3-D	

Essential	Optional
Sheets 56 and 57	Square tiles (about 60 for each pair of pupils)
Centimetre squared paper, scissors	Tracing paper
Multilink cubes (about 40 per group)	
Practice booklet pages 100 and 101	

A **Pentominoes** (p 233)

Optional: square tiles, tracing paper

While the main aim of this section is to enable pupils to recognise different nets in later work, a range of other skills and ideas can be developed, such as congruence, and these are described below.

◊ Pupils could start with shapes made from three squares and then four squares (trominoes and tetrominoes).

◊ After showing pupils what a pentomino is, ask them to find all possible different pentominoes. There is likely to be some discussion on the possible meanings of 'different' and 'same'. 'Different' is usually taken to mean non-congruent. Tracing paper helps pupils identify pentominoes that are the same. Tiles are useful to make the pentominoes.

The 12 pentominoes are shown below with possible labels.

'They all worked hard to find the twelve pentominoes – a Mars bar was at stake! The concept of same shape/different position came slowly to some but eventually they all understood.'

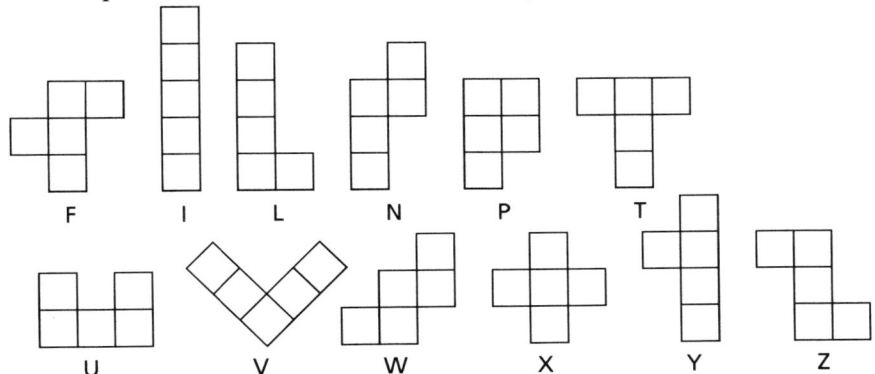

F   I   L   N   P   T

U   V   W   X   Y   Z

◊ Ask pupils how they can be sure they have found them all. Encourage them to be systematic. For example, they could find all those with five squares in a row, then four in a row, then three in a row and finally two in a row.

◊ When pupils think they have found all 12 pentominoes they could compare their results with the set on sheet 56.

◊ An additional activity is to fit all twelve pentominoes together to make a rectangle with no gaps.

There are 3719 solutions but don't expect any to be found quickly! The number of solutions for each possible size of rectangle is

6 by 10   2339 solutions
5 by 12   1010 solutions
4 by 15   368 solutions
3 by 20   2 solutions (very difficult)

◊ One school used the pieces as the basis of work on area, perimeter and symmetry. The perimeter of each shape was found and the shapes with the smallest and largest perimeters noted. The shapes were then classified as having reflection or rotation symmetry. The pupils found it useful to be able to physically turn and fold the shapes.

The 8 by 8 game (p 233)

This uses a set of pentomino pieces, and gives opportunities for strategy.

Sheets 56 and 57 (one of each for each group)

◊ The game is best played in groups of two or three.

◊ Encourage pupils to consider strategies for winning. For example, with two players the second player should try to move so that there is room only for an even number of pieces.

◊ One extension activity is to try to place all 12 pentominoes on the 8 by 8 board with four squares left over. This is difficult but solutions exist for all positions of the four squares.

◊ Another extension is to try to place pentominoes on the board in such a way that none of the rest can be added. The minimum number is five.

B Nets (p 234)

◊ Some pupils may need to cut out the pentominoes from squared paper to confirm which of them make an open cube. Encourage pupils to look at a possible net first and try to imagine it folding before cutting it out.

Design a poster (p 234)

The eight possible designs for a net of an open cube from the list of the 12 pentominoes in section A are F, L, N, T, W, X, Y, Z.

C Seeing in 3-D (p 235)

Multilink cubes (about 40 per group)

Four cubes (p 235)

This activity is best carried out in small groups. Pupils make their own sets of shapes from four cubes and then discuss whether they are indeed different from those made by others in the group. Pupils could record their results with a digital camera and produce a print for wall display or for their own record.

These are the eight distinct shapes using four cubes.

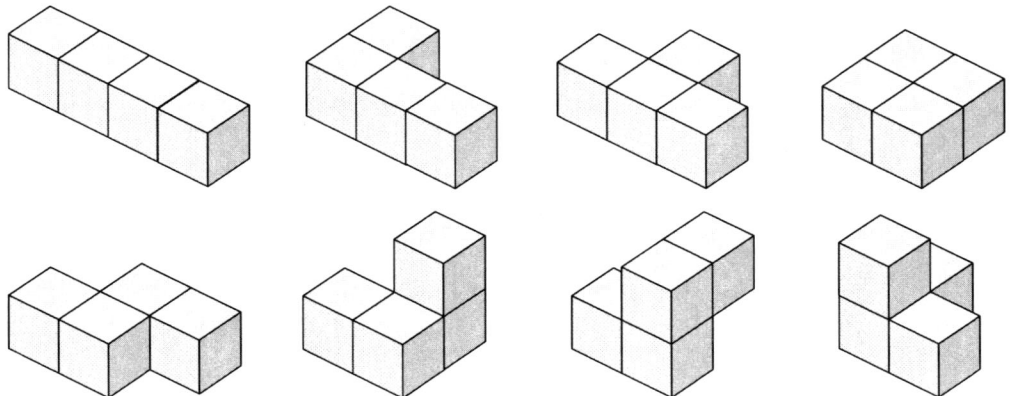

You could extend this by asking pupils to find distinct shapes with five cubes. However, there are 32 shapes so 160 cubes are needed to make a whole set. Pupils could instead be asked to produce just 10 distinct shapes.

B Nets (p 234)

B1 (a) Yes   (b) Yes   (c) No

     (d) Yes   (e) No   (f) No

B2 (a) No   (b) Yes

     (c) No   (d) Yes

C Seeing in 3-D (p 235)

C1 A and D, B and H, C and F, E and G

C2 Q

C3 24

C4 (a) 36   (b) 60   (c) 400

C5 David   3 cubes
Susan   7 cubes
Spencer   11 cubes

C6 Peter   5 cubes
Ros   15 cubes
Meena   9 cubes
Rahim   20 cubes

C7 Letter L   2 cubes
Letter H   5 cubes
Letter O   8 cubes

What progress have you made (p 237)

1 All three are nets.

2 F

Practice booklet

Section B (p 100)

    1 (a) Yes   (b) Yes   (c) No

    2 (a) Yes   (b) Yes   (c) No

       (d) Yes   (e) No   (f) No

Section C (p 101)

    1 A and E; B and C; D and F; G and H

    2 (a) 30 cubes   (b) 160 cubes

    3 Graham   7 cubes
       Hasima   11 cubes
       Lucy   17 cubes

㊶ **Ratio**

This unit is a basic introduction to the idea of ratio, solving simple problems using informal methods.

Practice booklet page 102

Ⓐ **Recipes** (p 238)

◊ You could begin by asking some simple questions to get pupils looking at the recipes, for example: 'How much sugar do you need for the apple crumble?' 'Which recipes need butter?'

Typical questioning can begin with 'Suppose you are going to cook sheek kebabs for 8 people.' Then (going round the class) '... how much mint would you need?', '... how much minced beef would you need?'

You can then ask about the amount of each ingredient required for these (roughly in order of increasing difficulty) and others of your own.

Topping for 2 pizzas, 3 pizzas, ...

20 scones, 30 scones, ...

10 pancakes, 15 pancakes

2 jam sandwich cakes

Sheek kebabs for 2 people, 1 person

Apple crumble to serve 3 people

You could ask diet-conscious questions such as: 'How much butter is there in one scone?' 'How much cream is there in one serving of pumpkin soup?', 'If the jam sandwich cake serves 5 people, how much sugar is there in each serving?'

Ⓑ **Making and sharing** (p 240)

Ⓐ Recipes (p 238)

A1 (a) 6 onions   (b) 120 grams

A2 (a) 15 eggs   (b) 350 grams

A3 (a) 150 grams   (b) 600 grams

     (c) 150 grams   (d) 500 ml

A4 (a) 20 grams   (b) 80 grams

     (c) (i) 20 grams (ii) 100 grams

     (d) 300 ml

A5

> **Apple crumble** *serves* **12** *people*
>   **6** large cooking apples
> **300** grams sugar
>   **2** pinches of cinnamon
> **400** grams flour
> **200** grams butter

A6 4 cakes

A7 40 pancakes

Ⓑ Making and sharing (p 240)

B1 (a) 1 litre   (b) 2 litres   (c) 4 litres

B2 (a) 3 litres lime cordial and
       6 litres lemonade

     (b) 12 litres

B3 Ben gets 30 stamps.

B4 (a) (i) 6   (ii) 15   (iii) 18

     (b) (i) 8   (ii) 20   (iii) 16

What progress have you made? (p 240)

1 (a) 30 jars   (b) 9 kg

2 10 kg gooseberries, 15 kg sugar,
   5 litres water

Practice booklet

Sections A and B (p 102)

1 (a) (i) 1 tin   (ii) 3 onions

     (b) 6 tins tomatoes
        18 medium onions
        12 teaspoons tomato paste
        6 teaspoons mixed herbs

2 (a) (i) 10 g   (ii) 80 g

     (b) 120 grams jam
        120 grams plain flour
        60 grams margarine

3 (a) 4 litres orange juice and 4 litres cola

     (b) 10 litres

4 (a) 15 cards   (b) 120 cards

     (c) 20 cards

Review 5 (p 241)

1 (a) 4.24   (b) 4.33
   (c) 5.58   (d) 0.08

2 Gareth is taller by 59 centimetres.

3 1 m 1 cm, 1.09 m, 1.10 m, 128 cm, 1.5 m

4 (a) 234.57   (b) 234.66
   (c) 244.56   (d) 334.56

5 PUZZLING

6 (a) 90 × 3   (b) 80 ÷ 4
   (c) 12 × 6   (d) 12 − 4

7 (a) None   (b) 15
   (c) (i) 1 p.m.   (ii) It was lunchtime
   (d) 6 p.m.

8

No of birds	Tally	Frequency
0–9	////	4
10–19	-H+/ /	6
20–29	-H+/	5
30–39	-H+/	5
40–49	-H+/	5
50–59	///	3
	Total	28

The pupil's bar chart

9

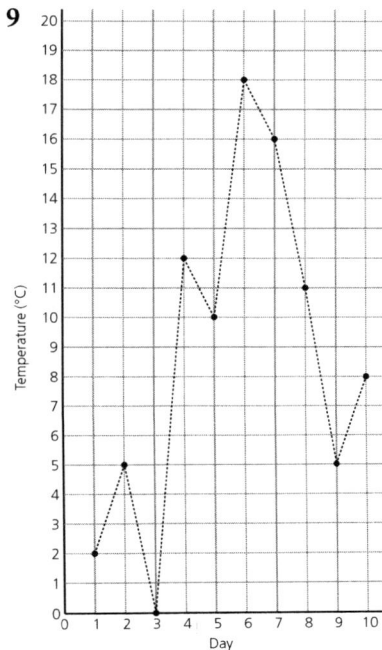

10 (a) 6   (b) 15   (c) 8   (d) 6
    (e) 7   (f) 5   (g) 10   (h) 6
    (i) 12   (j) 18

11 Harriet is 6 years old and 100 cm tall.

12 A, B, D, E and F can be.

13 Q is the odd one out.

14 (a) (i) 200 g   (ii) 150 g   (iii) 6
    (b) (i) 100 g   (ii) 50 g
    (c) (i) 20 g   (ii) 160 g
    (d) 600 g

15 (a) 740   (b) 344   (c) £3.11
    (d) £1.45   (e) 1944   (f) 1802
    (g) 5 rem 5 (h) 147 rem 1

16 (a) 6.3   (b) 4.6
    (c) 15.2   (d) 14.6

17 (a) 84p   (b) £336
    (c) 7p   (d) £4

18 (a), (c), (e)

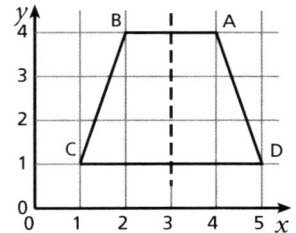

    (b) A (4, 4)   B (2, 4)   C (1, 1)
    (d) AB and CD
    (f) 3.2 cm   (g) 12.4 cm

19 (a) 4880   (b) 4900   (c) 5000

20 (a) $5\frac{3}{4}$ hours or 5 hours 45 minutes
    (b) 13 degrees
    (c) $4\frac{3}{4}$ hours or 4 hours 45 minutes
    (d) 10 degrees
    (e) 7:05 p.m.

21 (a) $\frac{1}{3}$ red and $\frac{2}{3}$ yellow

   (b) $\frac{5}{9}$ red and $\frac{4}{9}$ yellow

   (c) $\frac{5}{10}$ or $\frac{1}{2}$ red and the same yellow

   (d) $\frac{2}{8}$ or $\frac{1}{4}$ red and $\frac{6}{8}$ or $\frac{3}{4}$ yellow

22 (a) 26, 30; the rule is add four.

   (b) 47, 44; the rule is subtract 3.

   (c) 1.7, 1.9; the rule is add 0.2.

   (d) 680, 710; the rule is add 30.

23 (a) Q   (b) P   (c) R

24 (a), (d) (half-size)

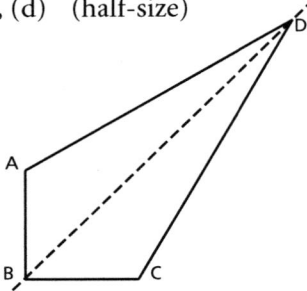

   (b) 8.2 cm   (c) 22.4 cm

25 (a) 12   (b) 18   (c) 4   (d) 12

   (e) 34   (f) 10   (g) 12   (h) 16

26 (a) 8   (b) 5   (c) 2   (d) 4

   (e) 20   (f) 10

Mixed questions 5 (Practice booklet p 103)

1 (a) 2.92   (b) 2.98   (c) 3.02

  (d) 0.87   (e) 0.96

2 (a) 0.06, 0.48, 0.5, 0.60, 0.68

  (b) 3.05, 3.4, 3.50, 3.54, 3.56

3 (a) $6 - 4$ (b) 5×2 (c) $24 \div 6$

4 (a) 24   (b) 12   (c) 30   (d) 100

5 (a) 8   (b) 18   (c) 32

6 (a) $2\frac{1}{3} = \frac{7}{3}$ (b) $3\frac{2}{5} = \frac{17}{5}$

  (c) $\frac{13}{3} = 4\frac{1}{3}$ (d) $\frac{17}{6} = 2\frac{5}{6}$

7 (a)

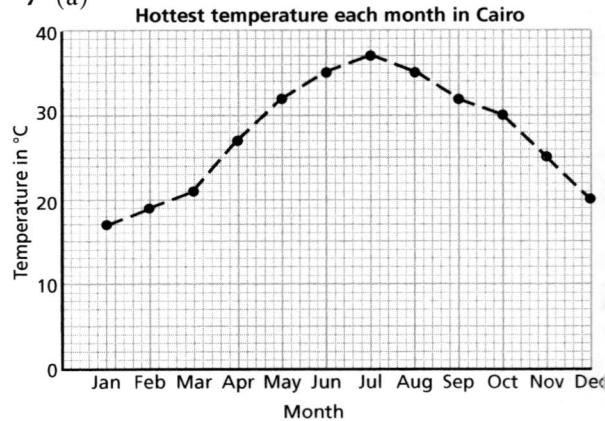

Hottest temperature each month in Cairo

  (b) January and February

8 View C

9 (a) 16 buckets of sand, 2 buckets of water

  (b) 22 buckets